INORGANIC REACTION CHEMISTRY:
SYSTEMATIC CHEMICAL SEPARATION

ELLIS HORWOOD SERIES IN ANALYTICAL CHEMISTRY

EDITORS: Dr. R. A. Chalmers and Dr. Mary Masson, University of Aberdeen

"I recommend that this Series be used as reference material. Its Authors are among the most respected in Europe". *J Chemical Ed., New York.*

Application of Ion-selective Membrane Electrodes in Organic Analysis
  G. BAIULESCU and V. V. COȘOFREȚ, Polytechnic Institute, Bucharest
Handbook of Practical Organic Microanalysis
  S. BANCE, May and Baker Research Laboratories, Dagenham
Ion-Selective Electrodes in Life Sciences
  J. COMER, MSE Scientific Instruments, Crawley, and D. B. KELL, University College of Wales, Aberystwyth
Inorganic Reaction Chemistry: Systematic Chemical Separation
  D. T. BURNS, Queen's University, Belfast, A. G. Catchpole, Kingston Polytechnic, A. TOWNSHEND, University of Birmingham
Reactions of the Elements and their Compounds
  D. T. BURNS, A. H. CARTER, A. TOWNSHEND
Quantitative Inorganic Analysis
  R. BELCHER, M. CRESSER, R. A. CHALMERS
Handbook of Process Stream Analysis
  K. J. CLEVETT, Crest Engineering (U.K.) Inc.
Automatic Methods in Chemical Analysis
  J. K. FOREMAN and P. B. STOCKWELL, Laboratory of the Government Chemist, London
Fundamentals of Electrochemical Analysis
  Z. GALUS, Warsaw University
Laboratory Handbook of Paper and Thin-Layer Chromatography
  J. GASPARIČ, Charles University, Hradec Kralove
  J. CHURAČEK, University of Chemical Technology, Pardubice
Handbook of Analytical Control of Iron and Steel Production
  T. S. HARRISON, Group Chemical Laboratories, British Steel Corporation
Handbook of Organic Reagents in Inorganic Analysis
  Z. HOLZBECHER et al., Institute of Chemical Technology, Prague
Analytical Applications of Complex Equilibria
  J. INCZÉDY, University of Chemical Engineering, Veszprém
Particle Size Analysis
  Z. K. JELINEK, Organic Synthesis Research Institute, Pardubice
Operational Amplifiers in Chemical Instrumentation
  R. KALVODA, J. Heyrovský Institute of Physical Chemistry and Electrochemistry, Prague
Atlas of Metal-Ligand Equilibria in Aqueous Solution
  J. KRAGTEN, University of Amsterdam
Gradient Liquid Chromatography
  C. LITEANU and S. GOCAN, University of Cluj
Titrimetric Analysis
  C. LITEANU and E. HOPÎRTEAN, University of Cluj
Statistical Methods in Trace Analysis
  C. LITEANU and I. RICĂ, University of Cluj
Spectrophotometric Determination of Elements
  Z. MARCZENKO, Warsaw Technical University
Separation and Enrichment Methods of Trace Analysis
  J. MINCZEWSKI et al., Institute of Nuclear Research, Warsaw
Handbook of Analysis of Organic Solvents
  V. SEDIVEČ and J. FLEK, Institute of Hygiene and Epidemiology, Prague
Methods of Catalytic Analysis
  G. SVEHLA, Queen's University of Belfast, H. THOMPSON, University of New York
Handbook of Analysis of Synthetic Polymers and Plastics
  J. URBANSKI et al., Warsaw Technical University
Analysis with Ion-selective Electrodes
  J. VESELÝ and D. WEISS, Geological Survey, Prague
  K. ŠTULÍK, Charles University, Prague
Electrochemical Stripping Analysis
  F. VYDRA, J. Heyrovský Institute of Physical Chemistry and Electrochemistry, Prague
  K. ŠTULÍK, Charles University, Prague
  B. JULAKOVÁ, The State Institute for Control of Drugs, Prague
Iso-electric Focusing Methods
  K. W. WILLIAMS, L. SODERBERG, T. LAAS, Pharmacia Fine Chemicals, Uppsala

INORGANIC REACTION CHEMISTRY: SYSTEMATIC CHEMICAL SEPARATION

D. T. BURNS
Professor of Analytical Chemistry
The Queen's University of Belfast

A. TOWNSHEND
Department of Chemistry
University of Birmingham

and

A. G. CATCHPOLE
Assistant Director
Kingston Polytechnic

ELLIS HORWOOD LIMITED
Publishers   Chichester

Halsted Press: a division of
JOHN WILEY & SONS
New York - Chichester - Brisbane - Toronto

First published in 1980 by
ELLIS HORWOOD LIMITED
Market Cross House, Cooper Street, Chichester, West Sussex, PO19 1EB, England

The publisher's colophon is reproduced from James Gillison's drawing of the ancient Market Cross, Chichester.

Distributors:

Australia, New Zealand, South-east Asia:
Jacaranda-Wiley Ltd., Jacaranda Press,
JOHN WILEY & SONS INC.,
G.P.O. Box 859, Brisbane, Queensland 40001, Australia.

Canada:
JOHN WILEY & SONS CANADA LIMITED
22 Worcester Road, Rexdale, Ontario, Canada.

Europe, Africa:
JOHN WILEY & SONS LIMITED
Baffins Lane, Chichester, West Sussex, England.

North and South America and the rest of the world:
Halsted Press, a division of
JOHN WILEY & SONS
605 Third Avenue, New York, N.Y. 10016, U.S.A.

British Library Cataloguing in Publication Data
Burns, Duncan Thorburn
    Inorganic reaction chemistry: Vol. I: Systematic Chemical Separation − (Ellis
    Horwood series in analytical chemistry).
    1. Chemistry, Inorganic   2. Chemical reactions
    I. Title   II. Townshend, A.
    III. Catchpole, Arthur George
    541'.39   QD151.2   79–42957
ISBN 0-85312-118-4 (Ellis Horwood Ltd., Publishers)
ISBN 0-470-26895-6 (Halsted Press)

Typeset in Press Roman by Ellis Horwood Ltd.
Printed in Great Britain by W. & J. Mackay Ltd., Chatham

List of Contents

Foreword. 9

Preface 13

Acknowledgments. 15

Chapter 1 Aspects of the History of Qualitative Analysis 17
    References. 26

Chapter 2 The Physico-Chemical Basis of Reactions in Aqueous Solution . . .29
  2.1   Types of Chemical Reaction in Aqueous Solutions. 29
  2.2   Principles of Chemical Equilibria 34
      2.2.1   Effect of Temperature and of Solvent 36
      2.2.2   The Significance of Activities. 37
  2.3   Applications of the Law of Mass Action. 38
      2.3.1   Reactions of Acids and Bases. .38
      2.3.2   Complexing Reactions 48
      2.3.3   Precipitation Reactions. 52
      2.3.4   Redox Reactions. 64
      2.3.5   Partition Equilibria and Liquid-Liquid Extraction 70
  2.4   Multiple Equilibria. 75
      2.4.1   Masking and Demasking 76
  2.5   Organic Reagents. 78
      2.5.1   Organic Complexing Agents. 80
  2.6   Visual Effects in Flames 87
  References. 88

Chapter 3 Systematic Inorganic Qualitative Analysis — A Theoretical
            Interpretation 93
     3.1   The MAQA Scheme 94
     3.2   Systematic Cation Separation Scheme .95
         3.2.1   Silver Group. 96
         3.2.2   Calcium Group 101
         3.2.3   Copper–Tin Group. 108
         3.2.4   Iron Group. 119
         3.2.5   Zinc Group 124
         3.2.6   Magnesium Group 126
         3.2.7   Anion Interferences in Group Separation of Cations. 127
     3.3   Systematic Scheme for the Separation of Anions. 128
     3.4   Non-Systematic Tests for Anions 137
     3.5   Systematic Analysis of Insoluble Substances. 138
         3.5.1   The Identification Scheme 141
     References 143

Chapter 4 Techniques and Apparatus for Semi-Micro Qualitative Analysis . . 147
     References 162

Chapter 5 Systematic Semi-Micro Qualitative Inorganic Analysis —
            Experimental Procedures 163
     5.1   Outline of Procedure for Examination of an Unknown Mixture . . . 163
     5.2   Analysis of Mixtures Containing Common Anions and Cations. . . . 164
         5.2.1   Preliminary Dry Tests. 164
         5.2.2   Preliminary Non-Systematic Examination for Anions 167
         5.2.3   General Tests 167
         5.2.4   Tests on the Original Material for Interfering Anions 170
         5.2.5   Preparation of the Sodium Carbonate Extract. 171
         5.2.6   Tests on the Sodium Carbonate Extract. 171
         5.2.7   Elementary Courses 173
     5.3   Group Separation Scheme for Cations 174
         5.3.1   Preparation of Solution 174
         5.3.2   Separation of the Cations into Groups. 176
         5.3.3   Silver Group. 179
         5.3.4   Calcium Group 180
         5.3.5   Parting of the Copper–Tin Group 182
         5.3.6   Copper Group. 183
         5.3.7   Tin Group 184
         5.3.8   Parting of the Copper–Tin Group (Alternative Procedure) . . . 186
         5.3.9   Iron Group. 187
         5.3.10   Zinc Group 188
         5.3.11   Magnesium Group 190

5.4 Group Separation Scheme for Anions 191
    5.4.1 Removal of Interfering Anions. 191
    5.4.2 Separation Scheme for Anions. 192
    5.4.3 Tests for Certain Combinations of Anions 197
5.5 Tests for Particular Substances. 201
5.6 Analysis of Mixtures also Containing Less Common Elements 202
    5.6.1 Preliminary Dry Tests. 202
    5.6.2 Preliminary Examination for Anions and Potassium 202
    5.6.3 Cation Analysis. 203
    5.6.4 Anion Analysis 210
5.7 Analysis of Insoluble Substances 213
References. 218

Chapter 6 Ring-Oven Technique in Qualitative Inorganic Analysis 221
    6.1 The Ring-Oven and its Operation 221
    6.2 Separations 224
    6.3 Systematic Separation Schemes 227
    6.4 Ring-to-Ring Separation 228
    6.5 Anions 229
    6.6 Electrographic Sampling. 229
    References. 231

Appendix List of Reagents. 233

Index 241

Foreword

Qualitative analysis, in some form or other, has been taught in advanced institutes of learning for at least two centuries. At one time it was the only means available for establishing the composition of materials, but this situation has gradually changed and today there is a host of alternative instrumental and other methods which can rapidly achieve the same purpose, often with considerably less labour. However, there are plenty of examples known where it may be more convenient to use the purely chemical method, for it can be quicker, simpler, cheaper and more informative. It is doubtful if one could argue the case for allotting a considerable part of laboratory time for training in these methods for these reasons alone, but qualitative analysis has a far more important pedagogic value. It acquaints the student with a great number of reagents and chemical reactions which he will encounter later in quantitative analysis; most of the separations which are used also have their quantitative counterparts, which have great importance in chemical analysis of various kinds. It provides the student, probably for the first time, with the chance to apply his knowledge of physicochemical principles at the bench; thus he encounters the effect of pH, buffer action, solubility products, complex formation, adsorption and other forms of precipitate contamination and so on [1]. When samples are analysed on the small scale, it trains the student in the technique of handling small amounts of material, a training which should be of use in other chemical operations.

During the last decade and even possibly before then, the amount of time devoted to qualitative analysis was reduced considerably, or eliminated completely. At one time it was taught intensively in schools in Britain, but no longer as far as is known. Whether it is important or not for a chemist to know anything about chemical reactions is a matter I do not propose to debate here, but we seem to be breeding a generation of chemists who have to go to a reference book to find a reaction that at one time most schoolboys would have known. This disease regrettably started in the U.S.A.; although we have reason to bless many of its exports, this particular one must rank with some of its other less desirable exports to Europe such as phylloxera, the grey squirrel and the concrete skyscraper. Happily, many of the leading Continental universities, especially those in the U.S.S.R. still recognize the great importance of this subject and it is still taught there extensively.

The great advances in other methods of establishing composition cannot be the main cause for the declining interest in qualitative analysis, for the other advantages listed previously must have been obvious to many teachers. In Britain, apart from a very small number of colleges well-known for the high standard of their analytical chemistry, the task of teaching qualitative analysis was often passed over to a junior member of the staff. To teach qualitative analysis effectively one needs to have not only an extensive knowledge of analytical reactions, but to have also the imagination and experience to cope with unpredictable effects and synergic phenomena. Unless it be taught by somebody with considerable experience and knowledge, the result would be chaos, as it often was.

Many teachers found the results of qualitative analysis disappointing and this was generally laid at the door of the combination of inexperienced staff and ham-handed students. Although this played a certain part, it was only a very minor one; the students were certainly not as ham-handed as was generally supposed. Most of the trouble lay with the inadequacies of some of the tests and separations. I have described [2] some of the bad workmanship that has gone into the qualitative tables, and my lecture entitled 'Myths and Legends in Analytical Chemistry' is still called for after a span of 20 years.

Even the physico-chemical interpretation of precipitation phenomena was done shoddily. Cobalt, nickel and zinc sulphides would be expected to precipitate in the acid sulphide group, according to the solubility products provided, the values for antimony and arsenic were always omitted, with no reasons given, and cadmium sulphide should theoretically have precipitated in concentrated hydrochloric acid. Almost every text book of qualitative analysis contained these inaccurate interpretations right up to the period when the decline started.

I realized long ago that there were so many problems in qualitative analysis that it was impossible for one man in a single lifetime to attempt to solve even a fraction of them. As a result the Midlands Association for Qualitative Analysis was formed in 1954; the general account of its origins has been described elsewhere [3]. Nevertheless, it is of interest to mention some of the problems that have been studied.

Some text books state that borate should be removed because calcium borate is insoluble in alkaline solution and would precipitate in the Iron Group. Other text books say that its removal is unnecessary because calcium borate remains soluble in ammonium chloride solution. Actually, it was shown that borate must be removed, but not for the reason given. The least soluble borates are those of zinc and cobalt and these are strongly adsorbed on hydrated aluminium oxide. Accordingly, if this particular combination occurs, zinc and cobalt can disappear when hydrated aluminium oxide is precipitated.

Even the tests described in the text books for the differentiation of such simple ions as carbonate and hydrogen carbonate are completely useless. Calcium salts are often recommended on the basis that calcium carbonate is insoluble and the hydrogen carbonate is soluble; however, under the general conditions of qualitative analysis, calcium carbonate is precipitated when calcium is added to a hydrogen carbonate solution. Magnesium salts are no better. Although there is no precipitate when magnesium ions are added to the hydrogen carbonate solution, when magnesium ions are added to a mixture of carbonate and hydrogen carbonate under the conditions of the analysis, the solution remains clear. Mercury(II) chloride is sometimes advocated as a reagent to differentiate the two anions. Carbonate yields a precipitate and hydrogen carbonate does not. However, the precipitate of basic mercury(II) carbonate is variously described as being white, yellow, orange, red-brown and red. In any case it is of no use for differentiating the two ions[4].

These and many other problems have been studied and solved in a series of papers published in *Mikrochimica Acta* which, at the time of writing, have reached the number of 44.

One of the decisions that the Committee had to settle before any work was started was the scale of operation. It was decided unanimously to use the semi-micro technique for this had all the advantages of speed, convenience and economy of small scale operations, whilst not demanding the great skill and experience required by micro-methods.

Semi-micro qualitative analysis had been pioneered in Britain by Mr. Harry Holness, then of the South-West Essex Technical College. He made several contributions to improving the separations and he designed much of the equipment that was later to be used by MAQA. It was mainly due to his efforts that a cheap centrifuge became available commercially.

It is now a quarter of a century since MAQA came into existence and some of the earlier members have passed away or left the Midlands area (this did not always bring about resignations; some members continued to travel to meetings in Birmingham from London, Manchester, North Wales, etc.). Altogether about 50 people have taken part in studying and refining the methods that have been used. When qualitative analysis was taught in the schools, the MAQA Elementary Tables were the most widely used.

It may seem paradoxical that a book of this kind should appear at a time when qualitative analysis is practised in only a few colleges in Britain, but it may well be that the tide will turn and some day it may have to be re-introduced. Certainly some of those institutions which were amongst the first to abandon this subject have since returned to it for the reasons stated earlier.

In conclusion, it should be mentioned that MAQA has been entirely self-supporting throughout its existence and it has often made grants to schools to promote the teaching of science. If there are any profits from the sale of this book they will also be devoted to the furtherance of analytical chemistry.

R. BELCHER
January, 1979

REFERENCES

[1] Stephen, W. I., *Educ. in Chem.*, 1969, **6**, 221.
[2] Belcher, R., *Analyst*, 1974, **99**, 802.
[3] Belcher, R., *Mikrochim. Acta*, **1956**, 1842.
[4] Osborne, V. G. and Freke, A. M., *Mikrochim. Acta*, **1964**, 179.

Preface

THE MIDLANDS ASSOCIATION FOR QUALITATIVE ANALYSIS

The Association was formed in 1954 (initially as the Midlands Analytical Chemistry Committee) on the initiative of Professor Ronald Belcher, who recognized the many deficiencies in the published systems of qualitative inorganic analysis and sought to remedy these by experimental investigation and discussion among a group of experienced analysts. The objects of the Association were to prepare recommended schemes for

(a) the systematic examination of common cations, based on the well-established classical procedure,

(b) the examination of anions,

(c) the examination of the less common elements,

(d) the examination of "insolubles".

The many investigations and testing of procedures by members have resulted in the publication, with continual revision, of an Advanced Scheme and an Elementary Scheme of Semi-Micro Qualitative Inorganic Analysis, a book of "Reactions of the Common Cations and Anions" and the preparation and distribution of an instructional film. Most of the developments have been reported in a series of 44 papers published in *Mikrochimica Acta*. This book represents a culmination of the work of the Association over the last twenty-five years. However, the members recognize that although the procedures described were regarded as the best available at the time of completing the manuscript, there remains room for further improvement and they do not regard the work on any of the procedures as the last word. The Association would be pleased to hear from any reader who feels he can contribute to further improvement. Comments or suggestions should be sent to The Secretary of MAQA, Dr. W. I. Stephen, Department of Chemistry, The University of Birmingham, P. O. Box 363, Birmingham B15 2TT.

MAQA

President:	Professor Ronald Belcher	(1958-)
Chairman:	Professor Ronald Belcher	(1954-1958)
	Dr. A. G. Catchpole	(1958-)
Secretary/Recorder:	J. A. Waddams	(1954-1958)
	Dr. W. I. Stephen	(1958-)
Recorder:	Dr. A. Townshend	(1964-1979)
Treasurer:	Dr. A. G. Catchpole	(1958-1959)
	Mr. O. B. Hayes	(1959-1968)
	Mr. V. J. Osborne	(1968-)

Other members who have actively contributed to the work of MAQA:

L. S. Bark	W. F. Jones
P. Brown	D. R. Lewis
D. T. Burns	H. Malissa
A. H. Carter	G. Marr
E. R. Clark	E. W. Moore
C. A. Colman-Porter	F. M. W. Olds
W. M. Dowson	G. F. Reynolds
G. A. Edwards	B. W. Rockett
G. E. P. Elliott	W. B. Shaw
D. L. Evans	H. C. Smith
P. R. Falkner	H. Weisz
P. G. W. Farr	T. S. West
J. W. Gregory	M. Williams
R. Harrison	J. Winterburn
Miss C. P. Jones	

Acknowledgments

The writing and preparation of this book owe much to the considerable advice and assistance received over many years from

A. H. Carter
E. R. Clark
W. M. Dowson
W. F. Jones
W. I. Stephen

Most of the members whose names have already been given have been involved in the work which has led to the publication of this book, either with original investigations, (the list of publications is given at the end of Chapter 5), extensive laboratory checking of both old and new procedures, especially by students, and also by the many informed technical discussions which have taken place over the years, and which have contributed to the theoretical interpretations included in Chapter 3.

The Association is also pleased to have had contributions from W. M. Campbell, Dr. T. S. West and Prof. H. Weisz which are included in Chapters 1, 2 and 3, and 6 respectively, and to acknowledge the editorial and chemical advice of Dr. R. A. Chalmers during the final stages of preparation of the book.

Almost all of the typing of the original draft manuscript was undertaken by Helen Catchpole, of intermediate drafts by Barbara Harrison and Peggy Bartram (Loughborough University) and the final draft by Sandra Geddis (Queens University of Belfast) to whom the Association owes a considerable debt of gratitude for so willingly undertaking so laborious and difficult a task.

The Association would especially like to record its indebtedness to Prof. Ronald Belcher for having the vision to initiate the MAQA and for the support and inspiration he has given to its activities throughout. Without these, the advances made in qualitative inorganic analysis by the Association would not have been realized and this book would not have been published.

Acknowledgments

The Association is also most grateful to Springer-Verlag, Vienna, for very kindly granting permission to reproduce an extensive amount of material from the 44 papers of the MAQA series in *Mikrochimica Acta*.

Finally, the authors wish to thank their wives for their forbearance.

D.T.B.
A.G.C.
A.T.

September 1979

Chapter 1

Aspects of the History of Qualitative Analysis

The MAQA scheme [1] is but one of a series of systems of qualitative inorganic analysis, some of which date back to the late eighteenth century. The earlier schemes are of interest in that many elegant separations and selective reactions, which remain in current qualitative and quantitative practice, originate from this work.

Inorganic qualitative analysis arose from the practical necessity of identifying useful ores and minerals. Consequently the art developed most rapidly where progressive metal mining industries flourished, pre-eminently in Sweden. Without injustice to other pioneer analysts, the Swedish chemist Torbern Bergman may be selected as the originator of systematic qualitative analysis. Probably the first texts of qualitative analysis recognizable as such are Bergman's two essays *De Analysi Aquarum* (1779) and *De Terra Gemmarum* (1780) [2]. Of the same period are *Réflexions sur l'analyse des pierres en général* (1779) by Vauquelin [3], and the remarkable series of papers by Klaproth which later made up the six volumes of his *Beiträge zur chemischen Kenntniss der Mineralkörper* (1795–1815) [4].

The value of analytical chemistry was recognized and commented upon by Parkes [5] in his *Essay on the Utility of the Study of Chemistry*, (1819): 'Is your son born to opulence, is he the heir to an extensive domain; make him an analytical chemist, and you enable him to appreciate the real value of his estate . . .'

In a later edition (1837) [6] he comments, '. . . By analysing the minerals which he discovers, he will ascertain . . . which of them may be worked with advantage.'

Rock analysis presents many difficulties, not least that of bringing the sample into solution. Bergman powdered minerals finely and mixed them with soda ash (sodium carbonate). The mixture was ignited strongly in an iron crucible for 3–4 hours, but after extraction with hydrochloric acid, there was often an insoluble residue which might be silica or undecomposed mineral. Klaproth showed that valid analysis required the complete dissolution of the sample. He used caustic potash (potassium hydroxide) as fusion agent, and carried out the fusion in silver crucibles. It was necessary to make frequent, repeated, acid extractions of the crucible contents. It became apparent that the most important factor in obtaining complete dissolution was particle size, and much attention was devoted to pulverization. Thomson records a gloomy story of a student who was too lazy to pay proper attention to the fine powdering of his sample of chrysoberyl. He found that fusion and extraction were of no avail 'and gave up the analysis in despair' [7].

At this stage there was no reliable scheme of analysis which could be applied generally to all samples. Thus each analysis was approached in the spirit of research. Nevertheless, certain rules did emerge; Campbell and Mallen [8] have drawn up schematic separations from the somewhat diffuse instructions of Bergman and his followers. Two of the separation schemes are given in Tables 1.1 and 1.2. One of the earliest examples of detailed published tables is that given by Noad [9].

Table 1.1

Bergman's Scheme for the Analysis of Stones

Powder the sample. Fuse with Na_2CO_3. Extract with HCl. Filter.

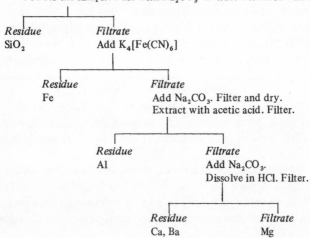

Glauber's account [10] of his 'miraculous salt' (1658), and the successful manufacture of Epsom salts (1695) [11], led to the revival of interest in the composition of mineral waters. Boyle [12] had posed a number of questions to help in classifying mineral waters, and had made use of a wide range of chemical reagents, including soap, syrup of violets, spirit of urine (ammonia), oil of vitriol (sulphuric acid), nitric acid and corrosive sublimate [mercury (II) chloride]. He also reintroduced the test-paper technique described by Pliny for detecting iron adulterants in verdigris.

Table 1.2

Klaproth's System for the Analysis of Stones

Powder the sample finely. Fuse with KOH. Extract with HCl. Repeat these operations until everything except SiO_2 has dissolved. Filter the HCl solution.

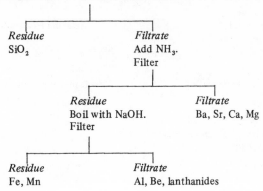

Residue	*Filtrate*
SiO_2	Add NH_3.
	Filter

Residue	*Filtrate*
Boil with NaOH.	Ba, Sr, Ca, Mg
Filter	

Residue	*Filtrate*
Fe, Mn	Al, Be, lanthanides

The systematic examination of mineral waters is first described by Bergman [2] in *De Analysi Aquarum* which discusses thoroughly and critically the colours and precipitates given by the commonly encountered metals and acids with certain reagents. Some of these are listed in Table 1.3. It is noteworthy [13] that the group of vegetable tinctures used by Bergman as acid–base indicators is only a small selection of those which have been used at various times. A reason for the proliferation of reagents in this class is that extracts in solution would not keep, and the coloured part of a plant would usually be available only during a short season. Not until 1864 did solid litmus appear in a stable form from which solutions could be made as required, although the related dyestuff cudbear had been manufactured in Scotland for a century.

Bergman also used tincture of galls as well as yellow prussiate of potash [potassium hexacyanoferrate(II)]. The non-specific nature of the gall test had been noted by Boyle [14], though some of his objections had little to do with mineral water analysis. If the reasonable assumption is made that iron is the only metal likely to be found in a mineral water apart from the alkali and alkaline earth metals, then tincture of gall remains a dependable reagent. Potassium hexacyanoferrate(II) became available in 1725 when Macquer [15] prepared it by digesting Prussian Blue with potassium carbonate solution. Made in this way, it usually contained some iron(III) ions and so regenerated Prussian Blue on acidification; it was therefore held in some suspicion when used as a test for iron. However, after Scheele had prepared the pure salt and Klaproth had shown how to prepare a solution for analytical use, the reagent increased in favour. The use of red prussiate of potash [potassium hexacyanoferrate(III)], prepared by Gmelin in 1822, as a test for iron(II) eliminated the necessity for allowing the mineral

Table 1.3
Bergman's Reagents for Mineral Water Analysis

Reagent*	Test for*
Tincture of heliotrope	
Tincture of Brazil wood	Acid and base (indicators)
Tincture of turmeric	
Syrup of violets†	
Tincture of galls	Iron
Potassium hexacyanoferrate(II)	Iron, copper, manganese
Sulphuric acid	Barium
Oxalic acid (and its potassium salts)	Lime
Ammonium sodium hydrogen phosphate (microcosmic salt)	
Potassium carbonate	All 'earths'
Ammonium carbonate	
Lime water	Carbon dioxide
Barium chloride	Sulphate
Calcium chloride†	Carbonate, but results ambiguous because sulphate may also be precipitated
Silver nitrate	Chloride and sulphide
Mercury(I) nitrate	To be used with caution because of
Mercury(II) nitrate	numerous reactions possible
Mercury(II) chloride†	Carbonate
Arsenic(III) oxide	Hydrogen sulphide
Lead acetate	Sulphide and sulphate (and chloride)
Potassium polysulphide/thiosulphate (liver of sulphur)†	Carbon dioxide
Iron(II) sulphate	Air and alkali
Alum†	Alkali, but not recommended
Alcohol	Salts
Soap	Hardness

*Modern nomenclature is used.
†Regarded as unnecessary.

water to stand until aerial oxidation had converted the iron(II) into iron(III).
A third test for iron, proposed by Gehlen [16], involved the formation of a
precipitate with ammonium succinate or benzoate. It was more in the nature
of a separation than an identification, and was so used by Thomson [7] to
separate iron from manganese.

The use of so many tests for elements such as iron was necessary because no test could be relied upon. This is not surprising in view of the rudimentary state of chemical theory at that time. Some workers even went so far as to question the utility of reactions with reagents in solution (commonly referred to as 'tests') in its entirety. Thus Kirwan [17] wrote in his *Essay on the Analysis of Mineral Waters* (1799): "Tests as commonly applied are generally allowed to afford only conjectural, not demonstrative, inference even of the species of salts contained in mineral waters, at least in the cases that most frequently occur".

The mistrust of 'tests' was attributable to the lack of any standard of purity or concentration. Reagents were usually made up in rain water, widely held to be the purest form of water obtainable, and in those instances where the solid salt was purchased through the ordinary commercial channels, gross adulteration was likely to be encountered [18]. Certain published tests can only be understood if the reagent is assumed to have been heavily adulterated. The growth of the chemical industry brought into being the new art of alkalimetry, requiring solutions of known concentrations, and this in turn provided the notion of the standard reagent [19].

On the other hand, Brande believed that yellow prussiate of potash was a universal reagent with which to identify the common metals [20]. He drew up a list of colour reactions of the reagent with various metals, from which it is apparent that a very high order of experience and colour perception would be required to make the test work. The precise distinction between 'yellowish white' and 'pale yellow' would not be easy to establish; moreover the stability of some of the precipitates would be a severely limiting factor.

The early analytical schemes for the examination of rocks and mineral waters made no provision for detecting the commercial metals such as copper, lead, antimony, zinc or nickel. To deal with these (later to be known as the Group II and Group IV metals) two techniques were evolved, namely hydrogen sulphide precipitation and the use of the blowpipe.

Hydrogen sulphide, long known as 'hepatic air' because it was made from liver of sulphur (impure potassium sulphide and polysulphide), was studied systematically by Scheele, who named it *'stinkende Schwefelluft'* in his *Experiments on Fire and Air* (1777) [21]. Scheele also showed how to prepare the gas from iron sulphide and acids. Bergman [2] in 1778 listed reactions of metal salts with an aqueous solution of hydrogen sulphide; the metals included silver, lead, zinc, copper, iron and arsenic. The fundamental paper, however, is that of Berthollet (1798), *Observations sur l'Hydrogène Sulphuré* [22], in which he describes carefully the precipitates and colours given by fourteen metals in acid or alkaline solution. Gay-Lussac (1811) [23], in a paper entitled *Sur la Précipitation des Métaux par l'Hydrogène Sulphuré*, showed that zinc, manganese, cobalt and nickel were precipitated in the presence of ammonia or weak acids such as tartaric or oxalic, though not in the presence of the mineral acids. These are of course the Zinc Group metals. Crosland [24] has pointed out that this paper was a decisive step on the way to the establishment of group separations.

The resulting demand for hydrogen sulphide gave rise to a series of generating devices [25]. The 1866 catalogue of the firm of J. J. Griffin lists twenty such pieces of apparatus, apart from Woulfe's bottles and paired aspirators. Kipp, the scientific instrument maker of Delft, whose firm is still in business, invented the famous apparatus which bears his name, in about 1860.

The detection of metals by means of the blowpipe perhaps merits a more extended discussion; its special virtue was that it could be used on completely unknown samples without previous classification or separation. The blowpipe, previously employed by jewellers and enamellers for centuries, was used for quasi-analytical work during the seventeenth century, as references by Hooke and Kunckel indicate. The use of the instrument became systematic in the middle of the eighteenth century. Cramer gave a careful description of it in 1741, and Cronstedt used it to discover nickel in 1751. Cronstedt's instructions for the use of the blowpipe initiated a spate of manuals for self-instruction in the technique.

The most difficult part of the blowpipe technique was the art of breathing in through the nose and out through the mouth at the same time so as to maintain a steady blast of air for some minutes. Several mechanical devices appeared which purported to make the management of the blowpipe easier; these included foot-bellows, metal reservoirs and even a bladder to be squeezed between the knees. Berzelius [26] rightly observed that it was easier to learn mouth-blowing than to manipulate such contrivances.

At first, much was written about the selection of the proper woods for making charcoal for supporting and reducing the sample; the Swedes preferred pinewood and the English chose the alderwood from which charcoal for gunpowder was made. Griffin's invention in 1843 of the prepared charcoal block, made from powdered charcoal, sodium carbonate, borax and rice flour, put an end to these discussions [27].

The source of the flame occupied the attention of many who used the blowpipe before the invention of the Bunsen burner. Rival claims were made for tallow and wax candles, and even the nature of the wick was solemnly argued. Portable oil lamps made their appearance in the 1830s. Griffin suggested a mixed fuel consisting of alcohol, turpentine and a few drops of ether, but it is difficult to believe that this ever became popular. The extent of the argument over apparently trivial details does, however, reveal much more about the difficult conditions in which the pioneer work on qualitative analysis was done.

There was clearly a sizeable amateur blowpipe cult because pocket kits of apparatus and chemicals, of varying degrees of sophistication, were offered for sale at prices to suit all pockets: collections of minerals for practice or recreation were also available. There was also an extensive literature on blowpipe analysis. Once more Bergman [28] was to the fore: his essay *De Tubo Ferruminatorio* (1779) dealt with gold, silver, platinum, mercury, lead, copper, iron, tin, bismuth, nickel, arsenic, cobalt, zinc, antimony and manganese. Griffin (who was a learned chemist as well as a scientific supplier) devoted part of his *Chemical Recreations* (1838) [29] to blowpipe analysis, but probably the most widely used works were those of Berzelius [26], Plattner [30] and Brush [31]. Readers with a taste for the more bizarre areas of scientific literature will find much to amuse them in W. A. Ross's *The Blowpipe in Chemistry, Mineralogy, and Geology* (1889) [32]. As late as 1942 the blowpipe was described as an invaluable instrument [33]. Some early blowpipes are illustrated in Fig. 1.1.

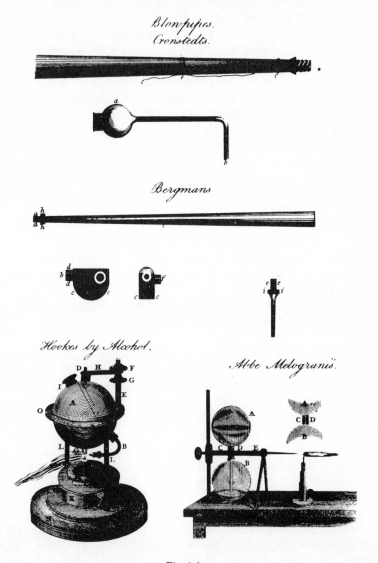

Fig. 1.1

In 1838, twenty years before the photochemical researches of Bunsen, Kirchhoff and Roscoe which led to spectroscopic analysis, Griffin published a table of flame colours [27]. In his later catalogues he offered wedge-shaped bottles to hold indigo solution to aid in distinguishing the flame colours given by lithium, sodium and potassium. Griffin recognized that tin, lead, antimony and arsenic caused destructive shortening of the platinum wire.

At this time qualitative analysis had an important part to play in the discovery of new minerals, which attracted the attention of Liebig, Berzelius and other leading chemists. This and much material in addition to that outlined herein is discussed by Szabadváry in his classic *History of Analytical Chemistry* [34]. The teaching of both qualitative and quantitative aspects of analytical chemistry in the United Kingdom before 1914 has been surveyed by Betteridge [35] and is discussed within the context of the growth of chemical teaching through the nineteenth century. More recently Stephen [36] has reviewed the place of qualitative inorganic analysis in undergraduate courses.

In Britain the setting up of the Department of Science and Art in 1854, with its examinations and its notorious 'payment by results' in 1859, led to a system of teaching which did much to degenerate qualitative analysis from a research tool into 'working through the tables'. This was criticised by the British Association, and in evidence given before the Royal Commission on Scientific Instruction in 1870 there were contemptuous references to 'test-tubing', a dull and mechanical routine carried out in a laboratory reeking of hydrogen sulphide. This carried over to spoil the image of chemistry itself, and even as late as 1926, Bishop Barnes of Birmingham recounted that "the old Greek play headmasters associated it not with culture but with low forms of manual dexterity and nauseous smells". Of particular influence between the two World Wars were the works of Treadwell [37] and of MacAlpine and Soule [38], the latter text being an invaluable source of early references.

The great early texts of Rose [39] and Fresenius [40] were followed at first by equally valuable works such as Galloway's *Manual of Qualitative Analysis* [41], packed with practical detail and descriptive inorganic chemistry, and calculated to last the student throughout his professional life. There arose, however, short manuals written in a kind of chemical shorthand which at times descended to esoteric jargon, which only the most gifted of teachers could bring to life.

The theory of dilute solutions, founded between 1885 and 1887 on the work of van't Hoff, Raoult and Arrhenius, was advocated by Ostwald as the first theoretical basis for qualitative analysis in his book *Wissenschaftliche Grundlagen der analytischen Chemie* (1894) [42, 43]. This was followed in 1917 by Stieglitz's *Elements of Qualitative Chemical Analysis* [44], the subtitle of which read "with special consideration of the application of the laws of equilibrium and of the modern theories of solution". T. B. Smith's *Analytical Processes — a Physico-chemical Interpretation* [45] did much to condition a new generation of students and teachers to a changed attitude to practical exercises in analysis.

Most historical divisions are in some way arbitrary, but it is probably fair to say that, after the theoretical enlightenment brought about by the works mentioned, the dawning of the modern period of qualitative analysis can be traced to two advances in technique. These were the scaling down of analytical procedures, and the introduction of organic reagents for inorganic ions.

Early examples of the detection of very small quantities of material are the Marsh test for arsenic (1836) and the Nessler test for ammonia in water (1856). Comprehensive schemes for inorganic microanalysis, however, are linked with the name of Emich, who in the 1890s began to develop the precipitation reactions under the microscope which Behrens had previously devised [46]. Behrens's monograph [47] outlines the contributions of previous workers, in addition to giving much original material. Later, Emich with his pupil Benedetti-Pichler devised apparatus for small-scale analysis [48]. The early history of microchemistry has been reviewed by Belcher [49].

Belcher and Wilson [50, 51] pioneered the teaching of inorganic microanalysis, including quantitative aspects, in the United Kingdom. Semimicro scale working was developed from the micro scale, the latter being too demanding in routine class usage. The first British book to include semimicro scale working was that of Vogel [52]. Holness [53] introduced various modifications to apparatus and to the parting of the copper-tin group with lithium hydroxide/potassium nitrate solution [54]. The MAQA has reviewed comparative schemes for the separation of ions within Groups, including the Iron [55], Silver [56], Copper-Tin [57] and Calcium Groups [58].

The use of organic reagents is rightly claimed as a modern development, but we need to remember that many of the reactions were known in the nineteenth century [59], in particular those of salicylic acid (1876), tannic acid (1881), 1-nitroso-2-naphthol (1884), pyrogallol (1887), fluorescein (1897), 2,2'-bipyridyl (1898), and formaldoxime (1899). Dimethylglyoxime came into use in 1905, dithizone in 1925, and 8-hydroxyquinoline (oxine) in 1927. The manufacture of organic reagents for analytical use became an important part of the fine chemical industry about 1930, the Hopkin and Williams *Organic Reagents for Metals* first appearing in 1933 [60]. The contributions of the fine chemical industry to the development of the subject by making pure reagents available should not be underestimated. The testing and preparation of pure reagents formed a large part of early experimental work, and some early texts [61, 62] could be regarded as the forerunners of the AnalaR Handbook *Standards for Laboratory Chemicals*.

Small-scale work and the use of specific reagents have been exploited to the full in the work of Fritz Feigl on spot-tests. In his *Qualitative Analyse mit Hilfe von Tüpfelreaktionen* (1931) [63] he introduced a large number of new reagents and also defined the loose term 'sensitivity' in two ways, as an identification limit and as a dilution limit. Both organic and inorganic analysis have benefited from this work [64, 65].

Thus qualitative analysis has progressed a long way from the analysis of stones, and has long ceased to deserve the old taunts of 'test-tubing'. It now encompasses large areas of inorganic, organic and physical chemistry, and possesses a coherent theory. It is able to illuminate both chemical education and professional practice, and to open up almost limitless fields for research. Bergman, whose vision was wide and whose grasp was firm, would have been delighted.

REFERENCES

[1] Belcher, R., *Mikrochim. Acta*, **1956**, 1842.
[2] Bergman, T., *Opuscula physica et chemica, pleraque antea seorsim edita, jam ab auctore collecta, revisa et aucta* . . . Holmiae, Upsaliae & Aboae, In Officinis librariis Magni Swederi, 1779-90.
[3] Vauquelin, L. N., *Ann. Chim.*, 1799, **30**, 81.
[4] Klaproth, M. H., *Analytical Essays towards Promoting the Chemical Knowledge of Mineral Substances* (trans. from German), Cadell & Davies, London, Vol. I, 1801, Vol. II, 1804.
[5] Parkes, S., *The Chemical Catechism, with Notes, Illustrations and Experiments*, 9th Ed., Baldwin, Cradock and Joy, London, 1819.
[6] Parkes, S., *A Catechism of Chemistry* . . . A new edition . . . considerably enlarged by W. Barker, London, 1837.
[7] Thomson, T., *Outlines of Mineralogy, Geology, and Mineral Analysis*, 7th Ed., Vol. II, Baldwin & Cradock, London, 1836.
[8] Campbell, W. A. and Mallen, C. E., *Proc. Univ. Durham Phil. Soc.*, 1959, **13**, *Series A (Science)*, 108.
[9] Noad, H. M., *Chemical Manipulation and Analysis*, Baldwin, London, 1852.
[10] Glauber, J. R., *The Works of John Rudolf Glauber, containing a great variety of choice secrets in the working of Metallic Mines and the separation of metals*, Packe & Milbourn, London, 1689.
[11] Grew, N., *Tractatus de Salis Cathartici amari* . . . Smith & Walford, London, 1695.
[12] Boyle, R., *Philosophical Works*, Vol. I, Innys and Osborn, London, 1725.
[13] Campbell, W. A. and Mallen, C. E., *Proc. Univ. Durham Phil. Soc.*, 1960, **13**, *Series A (Science)*, 168.
[14] Boyle, R., *Philosophical Works*, Vol. III, Innys and Osborn, London, 1725.
[15] Macquer, P. J., *A Dictionary of Chemistry* (trans.), 2nd Ed., 3 Vols. Cadell and Elmsly, London, 1777.
[16] Gehlen, see [7], p. 378.

[17] Kirwan, R. *An Essay on the Analysis of Mineral Waters*, Myers, London, 1799.

[18] *Journal of the House of Commons*, 1747, **25**, (23 March), 595.

[19] Henry, W., *Elements of Experimental Chemistry*, 11th Ed., Vol. II, p. 542, Baldwin & Cradock, London, 1829.

[20] Brande, W. T., *Manual of Chemistry*, 2nd Ed., Vol. II, p. 126, Murray, London, 1821.

[21] Scheele, C. W., *Chemical Observations and Experiments on Fire and Air*, (trans. J. R. Forster), Johnson, London, 1780.

[22] Berthollet, C. L., *Ann Chim.*, 1798, **25**, 233.

[23] Gay-Lussac, J. L., *Ann. Chim.*, 1811, **80**, 205.

[24] Crosland, M. P., *The Society of Arcueil*, p. 321, Heinemann, London, 1967.

[25] Aynsley, E. E. and Campbell, W. A., *J. Chem. Educ.*, 1958, **35**, 347.

[26] Berzelius, J. J. *Die Anwendung des Löthrohrs*, Schrag, Nürnberg, 1828.

[27] Griffin, J. J., *Proc. Glasgow Phil. Soc.*, 1843, *I*, 158.

[28] Bergman, T., (*De tubo ferruminatorio, ejusdemque in examinandis corporibus usu, commentatio*). *Abhandlung vom Gebrauche des Löthrohres, bey Untersuchhung der Mineralien*, 1779.

[29] Griffin, J. J., *Chemical Recreations*, 8th Ed., p. 148, Griffin, Glasgow, 1838.

[30] *Plattner's Manual of Qualitative and Quantitative Analysis with the Blowpipe*, revised and enlarged by T. Richter, edited by T. H. Cooksley, Chatto & Windus, London, 1875.

[31] Brush, G. J., *Manual of Determinative Mineralogy with an Introduction on Blowpipe Analysis*, revised by S. L. Penfield, Wiley, New York, 1898.

[32] Ross, W. A., *The Blowpipe in Chemistry, Mineralogy and Geology*, 2nd Ed., Crosby Lockwood, London, 1889.

[33] Read, H. H., *Rutley's Elements of Mineralogy*, 23rd Ed., Murphy, London, 1942.

[34] Szabadváry, F., *History of Analytical Chemistry*, Pergamon, Oxford, 1966.

[35] Betteridge, D., *Talanta*, 1969, **16**, 995.

[36] Stephen, W. I., *Educ. in Chem.*, 1969, **6**, 221.

[37] Treadwell, F. P., (trans. and revised by W. T. Hall), *Analytical Chemistry*, Vol. I, *Qualitative Analysis;* Vol. II, *Quantitative Analysis*, 8th Ed., Wiley, New York, 1932.

[38] MacAlpine, R. K. and Soule, B. A., *Qualitative Chemical Analysis*, Van Nostrand, New York, 1933.

[39] Rose, H., *Handbuch der analytischen Chemie*, 1831; *A Manual of Analytical Chemistry*, trans. J. J. Griffin, Tegg, London, 1831.

[40] Fresenius, C. R., *Anleitung zur qualitativen chemischen Analyse*, Bonn, 1841; *Elementary Instruction in Chemical Analysis*, trans. J. J. Bullock, Churchill, London, 1843.

[41] Galloway, R., *Manual of Qualitative Analysis,* Churchill, London, 1850.
[42] Ostwald, W., *Wissenschaftlichen Grundlagen der analytischen Chemie,* Engleman, Leipzig, 1894.
[43] Ostwald, W., *The Scientific Foundations of Analytical Chemistry,* trans. G. McGowan, 3rd English Ed., Macmillan, London, 1908.
[44] Stieglitz, J., *Elements of Qualitative Chemical Analysis,* The Century Co., New York, 1917.
[45] Smith, T. B., *Analytical Processes – a Physico-chemical Interpretation,* Arnold, London, 1929.
[46] Behrens, T. H., *Chem. News,* 1886, **54**, 196, 208, 220, 232, 243, 253, 266, 279, 289, 301, 316.
[47] Behrens, T. H., *A Manual of Microchemical Analysis,* Macmillan, London, 1894.
[48] Benedetti-Pichler, A. A., *An Introduction to the Microtechnique of Inorganic Analysis,* Wiley, New York, 1942.
[49] Belcher, R., *Proc. Anal. Div. Chem. Soc.,* 1975, **12**, 77.
[50] Belcher, R. and Wilson, C. L., *Qualitative Inorganic Microanalysis,* Longmans, London, 1946.
[51] Belcher, R. and Wilson, C. L., *Inorganic Microanalysis – Qualitative and Quantitative,* Longmans, London, 1957.
[52] Vogel, A. I., *A Text-book of Qualitative Chemical Analysis including Semimicro Qualitative Analysis,* 3rd Ed., Longmans, London, 1945.
[53] Holness, H., *Semi-micro Apparatus and Technique,* Pitman, London, 1953.
[54] Holness, H., *Advanced Inorganic Qualitative Analysis,* Pitman, London, 1957.
[55] Burns, D. T., *Mikrochim. Acta,* **1965**, 920.
[56] Burns, D. T., *Mikrochim. Acta,* **1967**, 147.
[57] Falkner, P. R. and Burns, D. T., *Mikrochim. Acta,* **1967**, 690.
[58] Jones, W. F., *Mikrochim. Acta,* **1967**, 1004.
[59] Belcher, R., *Pure Appl. Chem.,* 1973, **34**, 13.
[60] *Organic Reagents for Metals,* Hopkin & Williams, London, 1933.
[61] Mitchell, J., *Manual of Agricultural Analysis,* Simpkin, Marshall, London, 1846.
[62] Normandy, A., *The Dictionaries to the Chemical Atlas,* Knight, London, 1857.
[63] Feigl, F., *Qualitative Analyse mit Hilfe von Tüpfelreaktionen,* Akademische Verlag, Leipzig, 1931.
[64] Feigl, F. and Anger, V., *Spot Tests in Inorganic Analysis,* 6th Ed., trans. R. E. Oesper, Elsevier, Amsterdam, 1972.
[65] Feigl, F., *Spot Tests in Organic Analysis,* 7th Ed., trans. R. E. Oesper, Elsevier, Amsterdam, 1966.

Chapter 2

The Physico-Chemical Basis of Reactions in Aqueous Solution

Most of the reactions used in qualitative analysis involve the interaction of ions and molecules in aqueous solutions. The success of the analytical procedures which have evolved over many years depends on the proper sequential application of reactions selected for their efficiency in bringing about the separations and identifications required. The success of these reactions can often be explained quantitatively by the application of basic physico-chemical principles [1]. This chapter describes many of these principles and gives detailed examples of their application to a wide range of reactions and separations used in qualitative analysis. Some reactions, however, such as those involved in the thermal degradation of solids and in flame tests, are more difficult to describe quantitatively, and here are discussed only qualitatively. Additionally, quantitative treatment of reactions in solution is not possible if the relevant thermodynamic constants are unknown, as is often the case. A great deal of work remains to be done before a comprehensive collection of such constants can be available.

2.1 TYPES OF CHEMICAL REACTION IN AQUEOUS SOLUTIONS

Chemical reactions can be classified in many ways. The classification and order in which they are described below allows the quantitative treatment to be developed logically. It should be appreciated that many chemical processes involve more than one type of reaction, and that the quantitative interpretation of such processes can be complicated. The examples of the different types of reaction that follow merely indicate the overall stoichiometry of the reactions and the nature of the reaction products. Other parameters, such as reaction rates and equilibria, are discussed later.

(1) *Acid–Base Reactions*

According to Brønsted, acids are substances which tend to lose protons (hydrogen ions) and bases are substances which tend to take up protons. Acids are therefore proton-donors and bases are proton-acceptors:

$$\text{Base} + H^+ \rightleftharpoons \text{Acid}$$

An acid and a base related in this way are called a **conjugate pair**.

While there are more general definitions of acids and bases, these are not needed for the present discussion. An acid–base reaction, therefore, will be accompanied by a change in the hydrogen-ion concentration of the solution. For example, bases, which include hydroxide and acetate ions and ammonia, accept protons and hence remove them from solution by forming the conjugate acids:

hydroxide ion $OH^- + H^+ \rightleftharpoons H_2O$

acetate ion $CH_3COO^- + H^+ \rightleftharpoons CH_3COOH$

ammonia $NH_3 + H^+ \rightleftharpoons NH_4^+$

In the reverse reactions the acid gives up protons and hence generates them in forming its conjugate base.

Although these reactions are written above as if free hydrogen ions are the reacting species, in aqueous solutions these ions are hydrated, and are often described by the formula H_3O^+, although $H_9O_4^+$ would be more correct. Other ions are also hydrated; most metal ions, for example, are complexed by water molecules, e.g. $[Ni(H_2O)_6]^{2+}$. For clarity and simplicity, the co-ordinated water molecules are not referred to in the subsequent discussion. Nevertheless, because of their influence on ionic size and charge density, and possible involvement in the reaction mechanism, their presence and participation in reactions should always be borne in mind.

(2) *Complex-forming Reactions*

Many anions and molecules are able to use their lone pair(s) of electrons to co-ordinate with a metal ion. When they do so, they are said to act as **ligands**. The resulting compound formed between the ligand and the metal ion is called a **complex**. The simplest examples are those in which the complex is the only reaction product, as in the formation of diamminesilver(I) ions from silver(I) ions and ammonia:

$$Ag^+ + 2NH_3 \longrightarrow [Ag(NH_3)_2]^+$$

and of tetracyanonickelate(II) ions from nickel and cyanide ions:

$$Ni^{2+} + 4CN^- \longrightarrow [Ni(CN)_4]^{2-}$$

In these examples, the products are **complex ions**. Uncharged complexes can also be formed, most of which are only sparingly soluble in water, especially if the ligand is an organic species (see p. 78), as with tris(8-hydroxyquinolinato)-aluminium(III):

Complexes with organic ligands can often be made more water-soluble by incorporating solubilizing groups in the ligand such as sulphonic ($-SO_3H$) or hydroxy ($-OH$) groups (p. 74). Complexes in which more than one atom in a single ligand donates electron pairs to a metal ion, as in tris(8-hydroxyquino-linato)aluminium(III), are called **chelates**.

(3) Polymerization Reactions

Many chemical species are capable of reacting in aqueous solution with like species to form chains or rings. This process is called polymerization, and in inorganic chemistry is particularly prevalent among oxy anions, including phosphate, molybdate, vanadate, chromate, periodate and tellurate. For example, the dichromate ion, $[O_3Cr-O-CrO_3]^{2-}$, is formed in this way from chromate ions, CrO_4^{2-}:

$$H^+ + CrO_4^{2-} \longrightarrow HCrO_4^-$$

$$2HCrO_4^- \longrightarrow Cr_2O_7^{2-} + H_2O$$

If only one type of atom other than oxygen is involved, as in the dichromate ion, the anion is termed an **isopoly anion**. Other examples of isopoly anions are the various polyphosphate ions, such as

$$[O_3P-O-PO_3]^{4-}, \quad [O_3P-O-PO_2-O-PO_3]^{5-}, =$$

pyrophosphate   tripolyphosphate
ion ($P_2O_7^{4-}$)   ion ($P_3O_{10}^{5-}$)

trimetaphosphate
ion ($P_3O_9^{3-}$)

Two or more species may combine to form a mixed polymer, for example the formation of 12-molybdophosphate ions from molybdate and phosphate ions:

$$12MoO_4^{2-} + PO_4^{3-} + 24H^+ \longrightarrow [PO_4(MoO_3)_{12}]^{3-} + 12H_2O$$

12-molybdo-
phosphate ion

The acid formed from such an anion is termed a **heteropoly acid**.

(4) *Reduction-Oxidation (Redox) Reactions*

Redox reactions involve changes in the oxidation states of the reactants, brought about by the transfer of electrons, as in the oxidation of iron(II) ions by cerium(IV) ions:

$$Ce^{4+} + Fe^{2+} \longrightarrow Ce^{3+} + Fe^{3+}$$

and in the disproportionation of copper(I) ions:

$$2Cu^+ \longrightarrow Cu^{2+} + Cu(s)$$

The reactions of oxyanions also involve the participation of hydrogen ions, as in the oxidation of iodide by iodate to form iodine:

$$IO_3^- + 5I^- + 6H^+ \longrightarrow 3I_2 + 3H_2O$$

(5) *Precipitation Reactions*

The formation of a solid phase from a reaction in solution is called precipitation. Many precipitates are formed in aqueous solution

(a) by simple ionic interactions resulting in the formation of compounds of high lattice energy, such as barium sulphate:

$$Ba^{2+} + SO_4^{2-} \longrightarrow BaSO_4(s)$$

(b) by reactions of ions with organic compounds, for example the precipitation of nickel ions with dimethylglyoxime:

dimethylglyoximate ion

the protonation of benzoate ions to form benzoic acid:

$$H^+ + C_6H_5COO^- \longrightarrow C_6H_5COOH(s)$$

or the precipitation of sulphate ions by 4-chloro-4′-ammoniumbiphenyl
ions [2]:

$$SO_4^{2-} + 2H_3N^{+}\!\!-\!\!\bigcirc\!\!-\!\!\bigcirc\!\!-\!\!Cl \longrightarrow (H_3N\!-\!\bigcirc\!\!-\!\!\bigcirc\!\!-\!\!Cl)_2SO_4$$

Many precipitates, especially those involving organic species, are soluble in
organic solvents. If the organic solvent is not miscible with water, the
precipitate can be extracted from the aqueous phase into the organic
solvent by shaking them together. For example, copper(II) ions react with
diethyldithiocarbamate ions to form a sparingly soluble complex, which
can be extracted into chloroform:

$$Cu^{2+} + 2(C_2H_5)_2N\underset{\parallel}{C}S^- \longrightarrow [(C_2H_5)_2N\underset{\parallel}{C}S]_2Cu(s)$$

$$\qquad\qquad\qquad\overset{\parallel}{S} \qquad\qquad\qquad\qquad \overset{\parallel}{S}$$

$$\qquad\qquad\text{diethyl-} \qquad\qquad\qquad\Big\downarrow CHCl_3$$

$$\qquad\qquad\text{dithiocarbamate}$$

$$\qquad\qquad\text{ion} \qquad\qquad\qquad\qquad \text{brown solution}$$

(6) Volatilization Reactions

The formation of a gas or vapour, by simple ionic reactions, results in the
removal of the volatile product from the aqueous phase. The release of carbon
dioxide on acidification of a hydrogen carbonate solution is a familiar example:

$$H^+ + HCO_3^- \longrightarrow H_2O + CO_2(g)$$

More complicated reactions are involved in the dehydration of formic acid by
concentrated sulphuric acid:

$$HCOOH \xrightarrow{\ -H_2O\ } CO(g)$$

and in the formation of volatile silicon tetrafluoride or hexafluorosilicic acid
from silica and hydrofluoric acid:

$$SiO_2(s) + 4HF \longrightarrow SiF_4(g) + 2H_2O$$

$$SiO_2(s) + 6HF \longrightarrow H_2SiF_6(g) + 2H_2O$$

(7) Catalysed Reactions

Catalysts can be used in analysis to speed up slow but potentially useful
reactions. The reaction between cerium(IV) and arsenic(III),

$$2Ce^{4+} + As^{3+} \longrightarrow 2Ce^{3+} + As^{5+}$$

is catalysed by iodide or by osmium(VIII), and one of these catalysts is always added before a cerium(IV)–arsenic(III) titration is done. Such reactions can also be used for the detection and determination of the catalytic species. For example, copper(II) accelerates the disappearance of the violet colour that is produced initially by the reaction of iron(III) thiocyanate with thiosulphate. As little as 20 ng of copper can be detected by this catalytic effect [3].

(8) *Reactions Involving Organic Synthesis*

Reactions involving the formation of organic compounds with characteristics such as insolubility, colour or smell are often made use of in qualitative analysis. For example, the condensation of hydrazine with salicylaldehyde in aqueous solution is used as a test for hydrazine. The salicylaldehyde hydrazone formed precipitates from acidic media:

salicylaldehyde   salicylaldehyde
hydrazone

Note, however, that the aldehyde *must* be added to the hydrazine solution, or an aldazine is formed. Another example is the esterification of amyl alcohol by acetic acid to form amyl acetate, widely used as a test for acetate ions:

$$C_5H_{11}OH + CH_3COOH \longrightarrow CH_3COOC_5H_{11} + H_2O$$

amyl alcohol   amyl acetate

2.2 PRINCIPLES OF CHEMICAL EQUILIBRIA

In describing the different types of reaction outlined above, no account was taken of *reaction rate* nor of the possibility that the reactions do not go to completion, but reach a state in which both reactants and products are present.

All reactions proceed at finite rates. They not only proceed in the *forward* direction, reactants $\longrightarrow$ products, (as indicated in the equations above) but also, as the concentration of reaction products increases, in the opposite direction, the *reverse* reaction, products $\longrightarrow$ reactants, becoming increasingly important. Thus for any reaction between species A and B to give the products C and D:

$$A + B \longrightarrow C + D$$

the reverse reaction:

$$C + D \longrightarrow A + B$$

must also be considered. The situation is summarized by writing the reaction as an *equilibrium*:

$$A + B \rightleftharpoons C + D \tag{2.1}$$

The forward reaction, $A + B \longrightarrow + C + D$, has the rate V_F given by:

$$V_F = k_F a_A a_B \tag{2.2}$$

where k_F is a proportionality constant known as the rate constant, and a_A and a_B are the activities of A and B respectively. As will be seen later, activities are closely related to concentrations. Equation (2.2) is a mathematical expression of the **Law of Mass Action**, as originally stated by Guldberg and Waage [4].

Similarly, the reverse reaction, $C + D \longrightarrow A + B$, has a rate V_R given by:

$$V_R = k_R a_C a_D \tag{2.3}$$

These equations show that V_F is greatest at the beginning of the reaction because the activities of A and B in solution are greatest at this time. As the reaction proceeds, the activities of A and B decrease, so V_F decreases, but the activities of C and D increase, so V_R increases. Eventually the rates of both reactions become the same, and then

$$V_F = V_R \tag{2.4}$$

When this occurs, there is no further net change in the composition of the solution with time and the system has reached **equilibrium**. By combining Eqs. (2.2) - (2.4), an expression for the **equilibrium constant**, K, is obtained:

$$k_F a_A a_B = k_R a_C a_D \tag{2.5}$$

therefore $\qquad\qquad \dfrac{a_C a_D}{a_A a_B} = \dfrac{k_F}{k_R} = K \tag{2.6}$

Values for the equilibrium constants have been accurately determined for numerous systems [5-7].

The same principles can be applied to any reaction. For instance, for the more general reaction:

$$nA + mB \rightleftharpoons xC + yD \tag{2.7}$$

the expression for the equilibrium constant becomes

$$K = \frac{a_C^x \, a_D^y}{a_A^n \, a_B^m}$$

Note that the components written on the right-hand side of the reaction equation appear in the numerator of the equilibrium constant expression, and those on the left-hand side, in the denominator. The activities are raised to the power corresponding to the number of ions or molecules of the particular reactant appearing in the reaction equation*.

2.2.1 Effect of Temperature and of Solvent

The discussion above implies that the value of K is constant only under conditions in which k_F and k_R are constant, namely at constant temperature and pressure, and constant solvent composition. An increase in temperature increases the rate of reaction. Thus a change of temperature changes the rate constant for a given reaction. The quantitative relationship between rate constant and temperature is given by the Arrhenius equation:

$$k = Ae^{-E/RT}$$

where A is a constant, E is the activation energy for the reaction, R is the gas constant, and T is the absolute temperature. However, the relative changes in k_F and k_R with temperature are unlikely to be equal, so the value of K changes with temperature, which results in changes in the activities of the species in equilibrium. For example, heating an aqueous solution of cobalt(II) and chloride ions from 20° to 80° converts the pink, octahedral hexa-aquocobalt(II) ion into the blue, tetrahedral tetrachlorocobalt(II) ion:

$$[Co(H_2O)_6]^{2+} + 4Cl^- \rightleftharpoons [CoCl_4]^{2-} + 6H_2O \qquad (2.9)$$
$$\text{pink} \qquad\qquad\qquad \text{blue}$$

Similarly, a change of solvent usually changes k_F and k_R differently, so the value of K also changes. For example, the formation of the blue chloro-complex is favoured by the addition of ethanol or acetone to the reaction mixture.

---

*This arises because reaction (2.7) formally involves the interaction of n molecules or ions of A and m molecules or ions of B,

$$A + A + \ldots\ldots + B + B + B \ldots\ldots \longrightarrow$$

It is often erroneously assumed that $1/n$ of the activity of A reacts as the first A in the equation, and so on. This would mean that the rate of the reaction involving nA is proportional to $(a_A/n)^n$. In fact, however, as all the species A are indistinguishable, each is capable of reacting as the first, second or nth A, and the reaction rate is proportional to a_A^n. Similarly, for B the reaction rate is proportional to a_B^m.

2.2.2 The Significance of Activities

In very dilute solutions ($\leqslant 10^{-6}M$), ions are, on average, sufficiently far apart to have little influence upon each other. Thus they participate in chemical reactions, unaffected by the presence of the other ions in solution. Under these conditions their activities can be equated to their molar concentrations:

$$a_A = [A] \tag{2.10}$$

where [A] represents the molar concentration of species A.

In more concentrated solutions ($\geqslant 10^{-4}M$), ions are closer together and do influence one another so that, for example, the movement of a solvated ion through the solution is impeded by other ions. Ions of opposite charge (counterions) are attracted to it and form an ionic 'atmosphere' which resists the movement of the particular ion through the solution. The activity of an ion is thus less than would be expected from its concentration. In such circumstances, activity is related to concentration by the activity coefficient, f:

$$a_A = f_A [A] \tag{2.11}$$

The greater the charge on an ion, the greater is the influence of that ion on the other ions present with it in solution. In order to take into account the effect of total ionic concentration *and* charge on activity coefficients, a parameter known as the **ionic strength** (μ) of the solution is used [8]. It is defined as:

$$\mu = \tfrac{1}{2}\sum_{1}^{i} [i] z_i^2 \tag{2.12}$$

where i denotes an ionic species and z_i is the charge on that ion.

The dependence of the ionic strength on the square of the charge indicates the large effect produced by doubly and triply charged ions. For example, if a solution is $0.1M$ in barium chloride and $0.2M$ in hydrochloric acid,

$$\mu = \tfrac{1}{2}\left([Ba^{2+}]\,2^2 + [Cl^-]\,1^2 + [H^+]\,1^2\right)$$

$$= \tfrac{1}{2}\left((0.1 \times 4) + \{(2 \times 0.1) + 0.2\} + 0.2\right)$$

$$= 0.5M$$

whereas for $0.3M$ hydrochloric acid alone it would be only $0.3M$. The activity coefficient of an ion, f_i, in dilute solution, is related to the ionic strength by the equation:

$$- \log f_i = 0.51 z_i^2 \mu^{\frac{1}{2}} \text{ (at } 25^\circ) \tag{2.13}$$

This is the Debye–Hückel equation [9]. In practice, the activity coefficient for an individual ionic species cannot be measured. The mean activity coefficient $(f_\pm)$ for the ion and its oppositely charged partner can be evaluated by using the following modification of Eq. (2.13):

$$- \log f_\pm = 0.51 z_+ z - \mu^{\frac{1}{2}} \qquad (2.14)$$

where z_+ and $z-$ are the charges on the positively and negatively charged ions.

In even more concentrated solutions ($\geqslant 0.01 M$) this simple treatment is no longer adequate, and activity coefficients have to be calculated on the basis of more complicated expressions.

Activity coefficients decrease with increasing ionic strength up to about 1-2, but then increase as the ionic strength is increased further. For example, the activity coefficient for sodium bromide is 0.8 in a solution of ionic strength 0.3; it reaches its minimum of about 0.7 at an ionic strength of 1.0, and is about 0.9 at an ionic strength of 2.0. Molecules which carry no net charge are much less influenced by the effects of ions in dilute solutions. The activity coefficients of molecules in dilute solutions of electrolytes are therefore almost unity.

Many reactions in qualitative analysis are carried out under conditions of high ionic strength which, however, may vary between quite wide limits. Under such circumstances, little or no advantage is gained by applying activity coefficients in equilibrium calculations, and the use of concentrations gives an adequate account of the particular reaction. This is also true for relatively slow reactions, such as precipitations, if insufficient time is allowed for equilibrium to be attained. In the remainder of this chapter, therefore, concentrations are used in place of activities. However, the need to use activities in systems that are more precisely defined than those in qualitative analysis should always be borne in mind. It should also be realized that the 'background' or non-reacting species may have specific effects on the activity coefficients, and due allowance must be made for this (for example lithium and potassium may produce different effects even though present at the same concentration).

2.3 APPLICATIONS OF THE LAW OF MASS ACTION [10]

2.3.1 Reactions of Acids and Bases

Strong (mineral) acids such as hydrochloric and nitric are completely dissociated in dilute aqueous solution. For example, in the reaction between hydrogen chloride and water:

$$HCl + H_2O \longrightarrow H_3O^+ + Cl^- \qquad (2.15)$$

ionization is essentially complete, and no meaningful equilibrium constant can be measured. Most acids, however, are incompletely dissociated, so their equilibrium constants can be measured. Typical examples are acetic and benzoic acids and hydrogen sulphide. If a simple dissociation reaction is written for the weak acid, HX:

$$HX \rightleftharpoons H^+ + X^- \tag{2.16}$$

the equilibrium constant, K_a, is given by the expression

$$K_a = \frac{[H^+][X^-]}{[HX]} \tag{2.17}$$

and is called the **acid dissociation constant.**[†]

If a weak acid is the only solute present, the concentration of hydrogen ions produced from a nominal concentration, c, of the acid can be calculated as follows:

$$[H^+] = [X^-] \text{ and } [HX] = c - [H^+]$$

so from Eq. (2.17) $\qquad K_a = \frac{[H^+]^2}{c - [H^+]} \tag{2.18}$

from which $[H^+]$ can be calculated.

If the acid is very weak, dissociation is slight, so $[H^+] \ll c$, and Eq. (2.18) simplifies to

$$K_a c = [H^+]^2 \tag{2.19}$$

Strong bases, such as sodium and potassium hydroxides, are also completely dissociated in aqueous solution, but weaker bases, such as ammonia, take part in an equilibrium reaction with water:

$$NH_3 + H_2O \rightleftharpoons NH_4^+ + OH^- \tag{2.20}$$

These equilibria of weak bases with water can be treated similarly to the equilibria of weak acids with water, to give an expression for the dissociation of the weak base, K_b. For ammonia, $K_b = [NH_4^+][OH^-]/[NH_3] = 1.8 \times 10^{-5}$ mole/1 at $25°$. Note that molecules of ammonium hydroxide (NH_4OH) are not formed.

[†] It is sometimes convenient to consider the reaction as an association or protonation: $H^+ + X^- \rightleftharpoons HX$, so that the equilibrium constant given by the expression, $[HX]/[H^+][X^-]$ is an **acid association constant**. This gives the equilibrium constant the same form as the stability constant of a complex (p. 49), with a consequently more logical application in complexation reactions in which acid–base equilibria are involved. However, this convention will not be used in this text.

Not all acids dissociate in a single step to liberate just one hydrogen ion per molecule; some dissociate in a number of successive steps, each of which liberates a hydrogen ion. Such acids are termed **polyprotic**. For example, sulphuric acid liberates two hydrogen ions:

$$H_2SO_4 \longrightarrow H^+ + HSO_4^- \tag{2.21}$$

$$HSO_4^- \rightleftharpoons H^+ + SO_4^{2-} \tag{2.22}$$

The first stage is virtually complete in aqueous medium, but the hydrogen sulphate ion, HSO_4^-, is a weaker acid than H_2SO_4 so an equilibrium exists for the second dissociation step, as shown.

Oxalic acid is a weak acid with respect to both of its ionization steps:

$$\begin{array}{ccc} COOH & & COO^- \\ | & \rightleftharpoons & | & + H^+ \\ COOH & & COOH \end{array} \tag{2.23}$$

$$\begin{array}{ccc} COO^- & & COO^- \\ | & \rightleftharpoons & | & + H^+ \\ COOH & & COO^- \end{array} \tag{2.24}$$

Orthophosphoric acid (H_3PO_4) has three dissociation steps and ethylenediamine-N,N,N',N'-tetra-acetic acid, EDTA, which exists as a double zwitterion,

$$\begin{array}{ccc} HOOCCH_2 \diagdown \qquad \diagup CH_2COOH & & HOOCCH_2 \diagdown \qquad \diagup CH_2COO^- \\ NCH_2CH_2N & \rightleftharpoons & \overset{+}{N}HCH_2CH_2\overset{+}{N}H \\ HOOCCH_2 \diagup \qquad \diagdown CH_2COOH & {}^-OOCCH_2 \diagup \qquad \diagdown CH_2COOH \end{array}$$

EDTA

has four.

The hydrogen-ion concentration produced in aqueous solution by these acids can be calculated by algebraic treatment similar to, but more complicated than, that used for the simple acids. The use of computer programs [11] reduces the labour involved in carrying out such complicated calculations.

Nomenclature of Acid Dissociation Constants

Acid dissociation constants are represented generally as K_{ab}, where b is the total number of hydrogen ions that have been released in yielding the product from the fully protonated acid. Thus, the first dissociation constant for oxalic acid, Eq. (2.23), is designated K_{a1} and that for the equilibrium shown in Eq. (2.24) is designated K_{a2}. The equilibrium

$$H_3PO_4 \rightleftharpoons 2H^+ + HPO_4^{2-} \tag{2.25}$$

can be considered as composed of the two successive steps

$$H_3PO_4 \rightleftharpoons H^+ + H_2PO_4^-$$

$$H_2PO_4^- \rightleftharpoons H^+ + HPO_4^{2-}$$

and the equilibrium constant for reaction (2.25) is the product of the individual constants, i.e. $K_{a1} K_{a2}$. The subscript a is often omitted for simplicity.

Dissociation of Water
Water itself is slightly dissociated into hydrogen ions and hydroxide ions

$$H_2O \rightleftharpoons H^+ + OH^- \tag{2.26}$$

and the dissociation constant for this equilibrium is:

$$K = \frac{[H^+][OH^-]}{[H_2O]} \tag{2.27}$$

As the concentration of the water is essentially constant, it is possible to define another constant,

$$K_w = [H^+][OH^-] \tag{2.28}$$

called the ionic product of water, which has the value 1.0×10^{-14} mole^2.l^{-2} at $25°$. It should be remembered that the value of K_w changes with temperature. For instance, at $0°$ it is 10^{-15} mole^2.l^{-2}, but at $60°$ it is 10^{-13} mole^2.l^{-2} [12].

The Concept of pH
As hydrogen ions often have important effects on reactions and equilibria, it is essential to have knowledge of the hydrogen-ion concentration in a solution, even if that concentration is very small. Sørenson [13] devised a convenient logarithmic means of expressing the hydrogen-ion concentration. He defined an operator p such that

$$pH = -\log[H^+] \tag{2.29}$$

Thus the pH of $10^{-5}M$ hydrochloric acid is $- \log (10^{-5}) = - (-5) = 5$. On the pH scale, a tenfold reduction in hydrogen-ion concentration results in an increase of one unit in the pH value. Solutions which have hydrogen-ion concentrations greater than $1M$ have negative pH values. The system is easily extended to accommodate alkaline solutions. From Eqs. (2.28) and (2.29),

$$pH = - \log (K_w/[OH^-])$$ (2.30)

$$= - \log 10^{-14} + \log [OH^-]$$

$$= 14 + \log [OH^-]$$ (2.31)

$$pOH^\dagger = - \log [OH^-]$$ (2.32)

therefore $$pH + pOH = 14 \text{ at } 25°$$ (2.33)

Thus, for $10^{-3}M$ sodium hydroxide

$$pOH = 3$$

$$pH = 14 - pOH = 11$$

A neutral solution is defined as one in which the hydrogen-ion concentration is the same as the hydroxide-ion concentration, that is,

$$[H^+] = [OH^-]$$

$$K_w = [H^+][OH^-] = [H^+]^2 = 10^{-14} \text{ mole}^2.1^{-2}$$

$$[H^+] = [OH^-] = 10^{-7}M$$

$$pH = 7 \text{ at } 25°$$

An acidic solution has a greater concentration of hydrogen ions then hydroxide ions, and thus has a pH < 7; similarly, an alkaline solution has a pH > 7.

$\dagger pX = - \log [X]$ can be used for expressing the concentration of any species X; for example pM is often used to express metal-ion concentration, or pCl to express chloride-ion concentration. Dissociation constants can similarly be expressed; $pK = - \log K$.

Acid–Base Indicators [14]

There are weak organic acids and bases which have appreciably different colours when they are ionized and un-ionized, or when they are in different ionic states. These can be used to indicate changes in acidity or alkalinity in aqueous solution and are termed acid-base indicators. For example, a solution of un-protonated Methyl Red (Ind⁻) is orange-yellow whereas that of the protonated (Ind.H) is red:

$$(CH_3)_2 \overset{\overset{+}{H}}{N}\text{—}\langle\bigcirc\rangle\text{—}N{=}N\text{—}\langle\bigcirc\rangle\text{—}SO_3^- \longrightarrow$$

Ind.H (red)

$$(CH_3)_2 N\text{—}\langle\bigcirc\rangle\text{—}N{=}N\text{—}\langle\bigcirc\rangle\text{—}SO_3^- + H^+ \qquad (2.34)$$

Ind⁻ (orange-yellow)

The equilibrium constant is

$$K_{Ind} = [Ind^-][H^+]/[Ind.H] \qquad (2.35)$$

The colour intensity of a dilute solution is usually proportional to the concentrations and the molar absorbances of the coloured species. When the pH of a solution is such that both forms of an indicator are present, the colour observed is a mixture of the colours of the two components and depends on the concentration ratio of the components, which is determined by the pH. It is usually considered that a 10:1 ratio of one form of an indicator to the other is sufficient to prevent visual detection of the minor component, so the colour perceived is that of the major component. Therefore, as the pH of a solution containing an indicator is changed, a colour change is observed only between $[Ind.H] = 10[Ind^-]$ and $10[Ind.H] = [Ind^-]$. Thus, from Eq. (2.35) the colour change is observed when the hydrogen-ion concentration changes over the range from $[H^+] = K_{Ind}10[Ind^-]/[Ind^-] = 10K_{Ind}$ (acid colour) to $[H^+] = K_{Ind}[Ind^-]/10[Ind^-] = 0.1K_{Ind}$ (alkaline colour) i.e. from pH $= pK_{Ind}-1$ to $pK_{Ind}+1$.

The pH range over which an indicator changes from its acid to its alkaline colour is called the colour change interval or the transition range of the indicator and is usually given as $pK_{Ind} \pm 1$. For Methyl Orange, $pK_{Ind} = 3.4$, so the transition range is approximately from pH 2.4 to pH 4.4. The transition ranges for some common indicators, together with the observed changes of colour, are given in Table 2.1.

Table 2.1
Some Common Acid–Base Indicators [15]

Indicator	Colour change acid $\longrightarrow$ alkali	Transition pH range
Thymol Blue[†]	red $\longrightarrow$ yellow	1.2 – 2.8
Methyl Orange	red $\longrightarrow$ orange	3.0 – 4.4
Methyl Red	red $\longrightarrow$ yellow	4.2 – 6.3
Bromothymol Blue	yellow $\longrightarrow$ blue	6.0 – 7.6
Thymol Blue[†]	yellow $\longrightarrow$ blue	8.0 – 9.6
Phenolphthalein	colourless $\longrightarrow$ red	8.0 – 9.8
Thymolphthalein	colourless $\longrightarrow$ blue	9.3 – 10.5

†Thymol Blue has two colour transitions, one in acidic solution, the other in alkaline solution.

These observations refer only to two-colour indicators. For indicators which are colourless in one form, slightly different considerations apply, and if the criterion for the colour change is the first detectable appearance (or final disappearance) of the colour, the concentration of the indicator will affect the pH at which this occurs. Thus for a one-colour indicator (e.g. colourless in acid), if c is the minimum detectable concentration of the coloured form and C the total indicator concentration, the colour change will be perceived at a pH equal to pK_{Ind} when $C = 2c$, pK_{Ind} −1 when $C = 10c$, and pK_{Ind} −2 when $C = 100c$.

Acid–base indicators have two main applications. They can be used to indicate the end-point of appropriate acid–base titrations, and to determine the pH of a solution. For example, Methyl Orange is red at pH 3, orange at pH 4 and yellow at pH 5. Thus the pH of a solution between 3 and 5 can be estimated from the colour of this indicator in the solution. The colour may be compared either with that of the indicator in solutions of known pH or, more usually, with the colour chart found in books of pH indicator papers.

A single indicator covers only a small pH range, but a judicious mixture of indicators can give rise to progressive colour changes over a wide range of pH. A typical example [16] is a solution containing Methyl Orange, Methyl Red, Bromothymol Blue and phenolphthalein. It is red at pH 3, orange at pH 5, yellow at pH 6, green-blue at pH 8, blue at pH 9 and violet at pH 10. This mixture is called a universal indicator.

If the pH at the equivalence point of an acid–base titration falls within the transition pH range of a particular indicator, that indicator is suitable for indicating the end-point of the titration. Thus Methyl Orange would be suitable for a strong acid–weak base titration having a pH of 3.4 ± 1 at equivalence. In general, for titration of a weak acid with a strong base an indicator with $pK_{Ind} \sim pK_a + 2$ should be suitable. For weak bases titrated with strong acids a pK_{Ind} of $\sim pK_b$ −2 is recommended.

Salts of Weak Acids and Bases

Although the sodium and potassium salts of weak acids are essentially completely ionized in the solid form, in aqueous solution the anions react with water and are said to be hydrolysed. For example, acetate ions are hydrolysed to give an equilibrium with acetic acid and hydroxide ions:

$$CH_3COO^- + H_2O \rightleftharpoons CH_3COOH + OH^- \tag{2.36}$$

This reaction results in an increase in the hydroxide-ion concentration, and hence a decrease in the hydrogen-ion concentration, so a solution of sodium acetate is alkaline. As the equilibrium constant K_h', for the hydrolysis reaction is known, the exact pH of a solution can be calculated for any concentration of sodium acetate in water, as follows:

$$K_h' = \frac{[CH_3COOH]\,[OH^-]}{[CH_3COO^-]\,[H_2O]} \tag{2.37}$$

The concentration of water in a dilute solution is essentially constant, and can be included in the equilibrium constant as in the derivation of K_w. Thus:

$$K_h = \frac{[CH_3COOH]\,[OH^-]}{[CH_3COO^-]} \tag{2.38}$$

$$= \frac{[CH_3COOH]K_w}{[CH_3COO^-]\,[H^+]} = \frac{K_w}{K_1} \tag{2.39}$$

where K_1 is the acid dissociation constant of acetic acid. If the initial sodium acetate concentration is c, it follows from Eq. (2.36) that

$$[OH^-] = [CH_3COOH] \quad \text{and} \quad [CH_3COO^-] = c - [OH^-]$$

Then, from (2.38) $$K_h = \frac{[OH^-]^2}{c - [OH^-]} \tag{2.40}$$

and the hydroxide-ion concentration is obtained by solving this quadratic equation.

For weakly alkaline solutions, $[OH^-] \ll c$, so Eq. (2.40) can be approximated to give

$$[OH^-] = \left(\frac{K_w c}{K_1}\right)^{\frac{1}{2}} \tag{2.41}$$

and hence $$pH = \tfrac{1}{2}pK_w + \tfrac{1}{2}pK_1 + \tfrac{1}{2}\log c \tag{2.42}$$

Salts formed from weak bases with strong acids can be treated similarly. For example, ammonium chloride solutions are acidic. This arises because ammonium ions hydrolyse in aqueous solution to give hydrogen ions:

$$NH_4^+ + H_2O \rightleftharpoons NH_3 + H_3O^+$$

The pH of the solution can be calculated from an expression similar to that deduced for sodium acetate. The pH of weak acid–weak base salt solutions can be calculated approximately from a combination of these equations.

Buffer Solutions

Anions of weak acids are strong Brønsted bases, i.e. have a strong affinity for hydrogen ions. Thus when a small concentration of hydrogen ions is added to a solution containing such anions, most of the hydrogen ions are removed by the anions to form undissociated weak acid, so the change in hydrogen-ion content of the solution is small. Addition of a base to a weak acid results in the neutralization of the base, with little consequent change in pH, until most of the acid has been neutralized. Thus a solution of a weak acid together with its anions, in the form of a salt, is resistant to pH change irrespective of whether a small amount of acid or alkali is added. Such solutions are called **buffer solutions**. Similarly, weak bases and their salts also form buffer solutions.

The pH of a buffer solution depends on the relative amounts of weak acid (or base) and its salt present, and on the dissociation constant of the acid (or base) involved. For a solution containing a weak acid (HX) and its sodium salt (NaX), the pH can be calculated as follows.

If in the aqueous solution the hydrogen ions present are produced only by the dissociation of HX, then

$$[X^-] = c_S + [H^+]$$

where c_S is the initial concentration of the salt NaX, and

$$[HX] = c_A - [H^+] \tag{2.43}$$

where c_A is the initial concentration of the acid HX. Thus

$$K_1 = \frac{[H^+][X^-]}{[HX]} = \frac{[H^+](c_S + [H^+])}{(c_A - [H^+])} \tag{2.44}$$

or $$[H^+]^2 + [H^+](c_S + K_1) - K_1 c_A = 0 \tag{2.45}$$

from which $[H^+]$ is easily obtained.

If the acid is sufficiently weak and not too dilute, so that $[H^+]$ is very small compared with both c_S and c_A, Eq. (2.44) simplifies to:

$$K_1 = \frac{[H^+]\, c_S}{c_A} \qquad (2.46)$$

Thus

$$pH = pK_1 + \log \frac{c_S}{c_A} \qquad (2.47)$$

Equation (2.47), known as Henderson's equation, readily enables the change in pH on adding a strong acid (or alkali) to a buffer solution to be calculated.

For example, if 0.01 mole of strong acid is added to a litre of a buffer solution initially $0.1M$ in acetic acid and $0.1M$ in sodium acetate, the pH of the resulting solution may be calculated as follows. K_1 for acetic acid is 1.8×10^{-5} mole/1. The pH of the original buffer solution is given by use of Eq. (2.47):

$$pH = \log \frac{1}{1.8 \times 10^{-5}} + \log \frac{0.1}{0.1} = 4.74$$

When the strong acid is added, essentially all the added hydrogen ions are taken up by the acetate ions, so that the acetic acid concentration is increased by $0.01M$ and the salt (anion) concentration is decreased by the same amount. Thus c_A effectively becomes $0.11M$ and c_S $0.09M$, and the pH changes to

$$pH = 4.74 + \log \frac{0.09}{0.11} = 4.66.$$

Addition of 0.01 mole of hydrogen ions to this buffer solution has therefore resulted in an increase of only 0.4×10^{-5} mole of hydrogen ions in the solution. The same amount of acid added to the same volume of pure water would have increased the amount of hydrogen ions by 10^{-2} mole, which is 2500 times as great!

Buffer Capacity

The resistance of a buffer solution to change in pH when an acid or base is added to the solution is termed its **buffer capacity**. It is expressed mathematically by the **buffer index** [17]:

$$\text{Buffer index} = \frac{d[\text{Base}]}{d\text{pH}} \text{ or } \frac{d[\text{Acid}]}{d\text{pH}} \qquad (2.48)$$

Thus, the greater the buffer index, the greater will be the quantity of acid or base required to change the pH by a given amount. It should be noted that the expression for the buffer index is the reciprocal, or inverse, of the slope of an acid-base titration curve. Figure 2.1 shows that the buffer index is greatest when an acidic or basic species has been half-neutralized, that is, when the salt and acid (or base) concentrations are equal. The pH of such solutions [Eq. (2.47)] is equal to pK_1 (for weak monobasic acid buffers) or $pK_w - pK_b$ (for weak monoacidic base buffers).

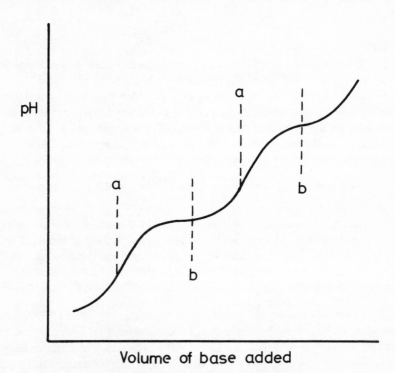

Fig. 2.1 – Change in pH during titration of a polyprotic acid with a base:
(a) equivalence points, (b) points at which $\dfrac{dpH}{d[\text{Base}]}$ is at a minimum and therefore buffer index is at a maximum.

The buffer capacity of a solution increases as the concentration of the buffer components increases. Thus if, in the previous example, the concentrations of acetic acid and acetate are each only $0.02M$, the initial pH will remain as 4.74. However, the addition of 0.01 mole of strong acid increases c_A to $0.03M$ and decreases c_S to $0.01M$, causing the pH to drop to 4.27. This is a much greater change than that for the more concentrated buffer solution.

2.3.2 Complexing Reactions
Reactions which involve the formation of complexes in solution can be

treated by simple equilibrium considerations. For example, the formation of the chlorocadmium(II) ion, $CdCl^+$ can be expressed by the equilibrium reaction:

$$Cd^{2+} + Cl^- \rightleftharpoons CdCl^+$$

The equilibrium constant for the reaction is given by the expression

$$k_1 = \frac{[CdCl^+]}{[Cd^{2+}][Cl^-]} \tag{2.49}$$

where k_1 is called the **stability or formation constant** of the complex.[†]

Cadmium forms further complexes containing up to four chloride ions:

$$CdCl^+ + Cl^- \rightleftharpoons CdCl_2$$

$$k_2 = \frac{[CdCl_2]}{[CdCl^+][Cl^-]} \tag{2.50}$$

$$CdCl_2 + Cl^- \rightleftharpoons CdCl_3^-$$

$$k_3 = \frac{[CdCl_3^-]}{[CdCl_2][Cl^-]} \quad \text{etc.} \tag{2.51}$$

k_1, k_2 and k_3 are called **stepwise** stability constants, because the reactions can be considered to occur in this fashion. Alternatively, such a sequence of reactions may be represented by one overall equation. For example, the formation of cadmium dichloride, or of trichlorocadmate(II) ions, may be considered as occurring in one step:

$$Cd^{2+} + 2Cl^- \rightleftharpoons CdCl_2$$

$$\beta_2 = \frac{[CdCl_2]}{[Cd^{2+}][Cl^-]^2} \tag{2.52}$$

$$Cd^{2+} + 3Cl^- \rightleftharpoons CdCl_3^-$$

$$\beta_3 = \frac{[CdCl_3^-]}{[Cd^{2+}][Cl^-]^3} \tag{2.53}$$

[†]Sometimes K is used instead of k.

where β_2 and β_3 are the **overall** stability or formation constants involving the addition of two and three ligands, respectively, to the metal ion. As with acid dissociation equilibria (p. 41), it can readily be shown that

$$\beta_2 = k_1 k_2 \tag{2.54}$$

$$\beta_3 = k_1 k_2 k_3 = \beta_2 k_3 \tag{2.55}$$

The nomenclature used for expressing stability constants should now be clear. The symbol k_n is used for the addition of a single ligand to an ion or a lower complex, where the subscript n indicates the number of ligands in the complex formed by the reaction. The β values are used for the addition of a number of ligands to the free or hydrated ion, the number of ligands bound being indicated by the subscript.

Other reactions can be treated similarly. For example, the overall stability constants for the diamminesilver(I) and the tetracyanonickelate(II) ions are given by the equations

$$\beta_2 = \frac{[Ag(NH_3)_2^+]}{[Ag^+][NH_3]^2} = 2 \times 10^7 \, l^2 \, mole^{-2} \text{ at } 25° \tag{2.56}$$

and

$$\beta_4 = \frac{[Ni(CN)_4^{2-}]}{[Ni^{2+}][CN^-]^4} = 10^{30} \, l^4 \, mole^{-4} \text{ at } 25° \tag{2.57}$$

The larger the value of a stability constant, the more strongly are the ligands bound to the central ion in the complex.

Provided the value of the stability constant is known, such expressions can be used for calculating the concentration of metal ions in a solution containing a complexing agent. The values of the stability constants for a large number of complexes have been measured, and tabulated values have been published [5].

For example, for a $10^{-4} M$ silver solution in $10^{-2} M$ ammonia, the free silver ion concentration at equilibrium is given by substituting the appropriate concentrations in Eq. (2.56). If it is assumed that practically all the silver is complexed, then $[Ag(NH_3)_2^+] = 10^{-4} M$, and the equivalent concentration of complexed ammonia is $2.0 \times 10^{-4} M$, so the concentration of free ammonia is $0.98 \times 10^{-2} M$. Thus, from Eq. (2.56)

$$[Ag^+] = \frac{1.0 \times 10^{-4}}{(0.98 \times 10^{-2})^2 \times 2 \times 10^7} = 5 \times 10^{-8} M$$

which agrees with the initial assumption that nearly all the silver ions are complexed.

Many ligands, such as oxalate, hydroxide, tartrate and 8–hydroxyquinolinate ions, are the anions of weak acids. With such ligands there is competition between the metal ions and hydrogen ions for bonding with the ligand, and both the complexing and the acid-base equilibria must be taken into account when calculating equilibrium concentrations. For example, for the formation of the lead acetate complex, CH_3COOPb^+, the equilibria involved are:

$$Pb^{2+} + CH_3COO^- \rightleftharpoons CH_3COOPb^+ \tag{2.58}$$

$$k_1 = [CH_3COOPb^+]/[Pb^{2+}][CH_3COO^-]$$

and $$H^+ + CH_3COO^- \rightleftharpoons CH_3COOH \tag{2.59}$$

$$K_1 = [CH_3COOH]/[H^+][CH_3COO^-]$$

These competitive reactions can be expressed by means of the equilibrium

$$Pb^{2+} + CH_3COOH \rightleftharpoons CH_3COOPb^+ + H^+ \tag{2.60}$$

The equilibrium constant for Eq. (2.60) is

$$*K = \frac{[CH_3COOPb^+][H^+]}{[CH_3COOH][Pb^{2+}]} \tag{2.61}$$

The symbol $*K$ for this type of equilibrium is a recommendation by IUPAC. It can readily be shown that

$$*K = k_1/K_1 \tag{2.62}$$

The extent of complex formation is then dependent on the acidity of the medium, as shown by Eq. (2.61). For a given ligand, the greater the stability of the complex with a metal ion, the lower the pH at which the complex can exist. This can be illustrated simply by considering the formation of the monohydroxo complexes of some metal ions. The hydroxide ion is the anion of a weak acid. The formation of a monohydroxo complex of a metal ion M^{2+} may be written as:

$$M^{2+} + OH^- \rightleftharpoons MOH^+ \tag{2.63}$$

$$k_1 = \frac{[MOH^+]}{[M^{2+}][OH^-]} \tag{2.64}$$

When 50% of the metal ions originally present are complexed,

$$[M^{2+}] = [MOH^+] \tag{2.65}$$

so
$$k_1 [OH^-] = 1 \tag{2.66}$$

and
$$[H^+] = k_1 K_W \tag{2.67}$$

Thus the pH at which there is 50% formation of a monohydroxo complex (neglecting the formation of higher complexes) is given by the equation:

$$pH = pK_W + pk_1 \tag{2.68}$$

Such pH values calculated for various hydroxo complexes are given in Table 2.2, together with the values of log k_1. They show that only the strongest complexes exist at the lowest pH values.

Table 2.2

Stability Constants [5] and pH of 50% Formation of Some Monohydroxo Complexes

Ion complexed	Ba^{2+}	Ca^{2+}	Mg^{2+}	Zn^{2+}	Fe^{2+}	Hg^{2+}
log k_1	0.6	1.5	2.6	4.2	5.7	10.3
pH for 50% complex formation	13.4	12.5	11.4	9.8	8.3	3.7

Most complexes form or dissociate sufficiently rapidly for equilibrium to be established; they are said to be labile. With others, reaction is too slow for this to happen, and such complexes are described as inert. For example, the dissociation of the hexacyanoferrate(II) ion, $[Fe(CN)_6]^{4-}$, in aqueous solution, with release of a cyanide ion, is very slow at room temperature, and many complexes of chromium(III) and cobalt(III) are particularly slow to dissociate even though the reactions concerned are thermodynamically favourable. For instance, $[Co(NH_3)_6]^{3+}$ can be kept almost indefinitely in $1M$ mineral acid, although equilibrium calculations show that less than 0.01% of the cobalt should be complexed in this way under such conditions. Care should be taken, therefore, to ensure that equilibrium conditions exist before applying calculations utilizing equilibrium data.

2.3.3 Precipitation Reactions

Many ionic reactions in aqueous solution result in the formation of a solid phase. When precipitation and dissolution reactions are reasonably rapid, equilibrium calculations can be applied to describe the process. If a typical precipitation reaction is represented by

$$A + B \rightleftharpoons AB \text{ (soln)} \rightleftharpoons AB \text{ (s)} \tag{2.69}$$

where $AB(s)$ denotes AB in the solid phase, the equilibrium constant is given by the expression:

$$K = \frac{[AB] \text{ (soln)}}{[A] [B]} \tag{2.70}$$

The concentration of AB molecules in the solution is known as the intrinsic solubility (p. 82). As this concentration is essentially invariant, it can be combined with K to give a new expression:

$$K_S = [A] [B] \tag{2.71}$$

where K_S is the **solubility product** for AB in a particular solvent, and is the product of the ionic concentrations of A and B that exist in the solution in equilibrium with the precipitate. This means that the solubility product is the ionic product for a saturated solution of AB.

In a saturated solution of AB, assuming complete dissociation into the components A and B,

$$[A] = [B] = \text{molar solubility of AB} = S_{AB}$$

$$K_S = [A]^2 = [B]^2 \tag{2.72}$$

$$S_{AB} = K_S^{\frac{1}{2}} \tag{2.73}$$

For a more general precipitation reaction, that of $A_n B_m$:

$$nA + mB \rightleftharpoons A_n B_m(s)$$

$$K_S = [A]^n [B]^m \tag{2.74}$$

$$[A] = nS_{A_nB_m} \qquad [B] = mS_{A_nB_m} \tag{2.75}$$

$$K_S = n^n m^m S_{A_nB_m}^{(n+m)} \tag{2.76}$$

$$S_{A_nB_m} = (K_S/n^n m^m)^{1/(n+m)} \tag{2.77}$$

Factors that Influence the Solubility of Precipitates [18]

(1) *The Common-Ion Effect*

The solubility product of a substance is a constant at a given temperature in a given solvent. Thus, for example, if more sulphate ions are added to a saturated solution of barium sulphate, the increase in sulphate-ion concentration is necessarily accompanied by a decrease in barium-ion concentration to maintain the constancy of the solubility product. Thus the solubility of barium sulphate, as measured from the concentration of barium ions in solution, is decreased by adding more of an ion common to the precipitate, namely sulphate. This effect is referred to as the **common-ion effect**.

From Eq. (2.73) a saturated barium sulphate solution is $10^{-5}M$ in both barium ions and in sulphate ions at $25°$, because $K_S = 10^{-10}$ mole^2.l^{-2}. If the sulphate ion concentration is increased to $10^{-2}M$ by adding sodium sulphate, the equilibrium concentration of barium is reduced to

$$[Ba^{2+}] = \frac{K_S}{[SO_4^{2-}]} = \frac{10^{-10}}{10^{-2}} = 10^{-8}M$$

Thus the solubility of barium sulphate in $10^{-2}M$ sodium sulphate solution is only a thousandth of that in water, at $25°$. However, this treatment is rather oversimplified because it neglects to take ion-pair formation into account, and the true solubility will probably be somewhat higher (see next paragraph).

(2) *Molecular Solubility of Ionic Precipitates*

It is generally assumed that the 'molecules' of an ionic precipitate have a negligible solubility. Thus, in Eq. (2.70), $[AB]_{soln}$ is not only assumed to be invariant but also to be extremely small. The latter assumption is often not true. The solubility of thallium(I) chloride in water, for example, arises not only from the release of thallium(I) and chloride ions into solution to give a $2 \times 10^{-2}M$ solution of each of these ions (p. 99), but also from solubility of Tl^+Cl^- ion-pairs, for which a value of $8 \times 10^{-4}M$ has been determined [19]; this accounts for 4% of the total solubility.

(3) *Organic Solvents*

Solubility products are usually measured in aqueous solutions. The presence of other solvents, however, has a marked effect on such constants. Ionic precipitates, such as barium sulphate, lead sulphate and silver chloride, are less soluble in solvents having a dielectric constant less than that of water. Thus, the addition of ethanol or acetone to aqueous solutions decreases the solubility of such precipitates. Lead sulphate, for example, is one tenth as soluble in 50% aqueous ethanol as it is in water. On the other hand, precipitates with organic reagents often show increased solubility in organic solvents. Typical examples are the enhanced solubility of bis(8-hydroxyquinolinato)zinc and bis(dimethylglyoximato)nickel in aqueous ethanol, compared to that in entirely aqueous media. In some instances the 'molecular solubility' of the precipitate is sufficient to make it desirable to include it in equilibrium equations.

(4) *The Formation of Soluble Complexes*

As already described, many metal ions form a series of complexes by stepwise addition of the ligand. Complexes in which the charge on the metal ion is balanced by the total charge on the ligands are usually only very slightly soluble in water. An example is mercury(II) iodide, HgI_2, which precipitates as a polymeric complex with the iodide ions acting as bridging ligands:

The stepwise addition of ligands can continue beyond the formation of the insoluble species; in the example above, the presence of a high concentration of iodide ions allows the formation of the soluble complex ions HgI_3^- and HgI_4^{2-} at the expense of the precipitate. Similarly, mercury(II) sulphide dissolves in the presence of a sufficient concentration of sulphide ions (this requires an alkaline solution, because H_2S is a weak acid) to form HgS_2^{2-}, and silver chloride dissolves in very concentrated chloride solutions as $AgCl_2^-$.

This effect is not limited to inorganic ligands. Thorium is precipitated by oxalate as $Th(C_2O_4)_2$ but the precipitate dissolves if excess of oxalate is added, the complex $Th(C_2O_4)_3^{2-}$ being formed. In such cases excess of ligand suppresses the solubility of the precipitate by the common-ion effect, but at the same time enhances the solubility by the formation of higher complexes, even to the extent of causing complete dissolution of the precipitate.

Likewise, species that form soluble complexes with the precipitated metal ion will compete with the precipitant for the metal ion. For example, cadmium and mercury(II) sulphides are appreciably more soluble in chloride ion solutions than in nitrate ion solutions because chloride ions form soluble chloro-complexes of the metal ions. In favourable instances ligands can be chosen to complex a given metal ion selectively, so that one metal ion can be precipitated without precipitation of another metal ion present with it in solution. This is an example of **masking** (p. 76).

(5) *Effect of pH*

When one of the ions involved in a precipitation reaction is the anion of a weak acid or the cation of a weak base, the hydrogen-ion concentration of the solution is important in determining how complete the precipitation is or, indeed, whether it occurs at all. For example, precipitation of calcium oxalate involves the oxalate ion, Ox^{2-} (i.e. $C_2O_4^{2-}$), which is the anion of fully dissociated oxalic acid, which is a weak acid:

$$Ca^{2+} + Ox^{2-} \rightleftharpoons CaOx \text{ (soln)}$$

$$CaOx \text{ (soln)} \rightleftharpoons CaOx.H_2O(s)$$

$$K_S = [Ca^{2+}][Ox^{2-}] \tag{2.78}$$

However, the oxalate ions will also take part in an equilibrium with hydrogen ions, and the concentration of *free* oxalate ions in Eq. (2.78) must also be the concentration of free oxalate ions dictated by the equilibria

$$H_2Ox \rightleftharpoons H^+ + HOx^- \qquad K_{a1} = [HOx^-][H^+]/[H_2Ox] \tag{2.79}$$

$$HOx^- \rightleftharpoons H^+ + Ox^{2-} \qquad K_{a2} = [Ox^{2-}][H^+]/[HOx^-] \tag{2.80}$$

The total concentration of oxalate species not combined with calcium is given by

$$[Ox]_{total} = [Ox^{2-}] + [HOx^-] + [H_2Ox]$$

$$= [Ox^{2-}]\{1 + [H^+]/K_{a2} + [H^+]^2/K_{a1}K_{a2}\} \tag{2.81}$$

The ratio $[Ox]_{total}/[Ox^{2-}]$ is obviously a function of the pH, and increases as the hydrogen-ion concentration increases. It is convenient to call this ratio $\alpha_{Ox(H)}$, the side-reaction coefficient for interaction of oxalate ions with protons (as indicated by the subscript). This notation, proposed by Ringbom [20], will be used frequently later in the text. The quantity α gives a quantitative estimate of the extent to which the oxalate is *not* available for precipitation (because it has been protonated). For a given total oxalate concentration the fraction present as free oxalate ions will be dependent on the pH and equal to $[Ox]_{total}/\alpha_{Ox(H)}$.

Thus Eq. (2.78) becomes

$$K_S = [Ca^{2+}][Ox]_{total}/\alpha_{Ox(H)} \tag{2.82}$$

Hence $$[Ca^{2+}] = K_S \alpha_{Ox(H)}/[Ox]_{total} \tag{2.83}$$

and for a fixed total oxalate concentration the calcium concentration and thus the solubility of the precipitate will increase as $\alpha_{Ox(H)}$ increases, i.e. as the pH is lowered. That is, less calcium oxalate is precipitated. This can be expressed by use of a conditional stability constant K'_S (so called because it takes a different value for each set of conditions, the value being constant if the conditions are constant[†]):

$$K'_S = [Ca^{2+}][Ox]_{total} = K_S \alpha_{Ox(H)} \qquad (2.84)$$

The main advantage of this system is that α can be calculated once and for all and expressed graphically as a plot of $\log \alpha$ *vs* pH (Fig. 2.2).

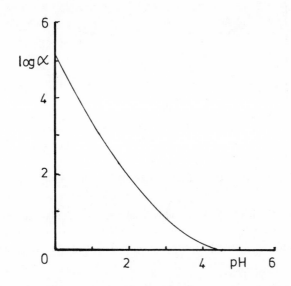

Fig. 2.2 – Change in $\log \alpha$ with pH for oxalic acid ($pK_1 = 1.4$, $pK_2 = 3.8$).

From Eq. (2.84),

$$pK'_S = pK_S - \log \alpha_{Ox(H)}$$

and pK'_S can be plotted as a function of pH simply by graphical subtraction of $\log \alpha$ from pK_S (which is independent of pH), as shown schematically in Fig.2.3.

† $[Ox]_{total}$ is often written as $[Ox']$ and is called the conditional oxalate concentration.

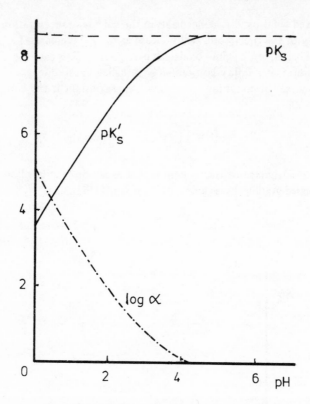

Fig. 2.3 – Graphical calculation of pK'_S ($pK_S = 8.6$).

It is then a simple matter to extend this treatment to calculate the pH at which K_S is exceeded, and the pH at which precipitation will be virtually complete, for any set of conditions. Typical results of such calculations for metal oxalates are shown in Table 2.3.

Table 2.3 – Solubility products of some metal oxalates, and the pH values at which the ionic concentration products are the same as the appropriate solubility products, and at which precipation is at least 99.9% complete.

M^{2+}	Pb^{2+}	Ca^{2+}	Sr^{2+}	Ba^{2+}	Mg^{2+}
K_S† [5]	3×10^{-11}	2×10^{-9}	5×10^{-8}	2×10^{-7}	9×10^{-5}
pH to attain K_S	−0.4	0.5	1.2	1.5	2.8
pH for 99.9% precipitation	1.1	2.0	2.7	3.0	4.3

$(K_{a1} K_{a2} = 2 \times 10^{-5}, [M^{2+}] = 10^{-3}M, [H_2Ox] = 10^{-2}M)$.

†Henceforth the units of constants will be omitted for simplicity, since it is easy to see what they should be.

(6) *Salt Effects*

A large concentration of a salt having no ions in common with a precipitate usually increases the solubility of the precipitate because in a medium of such high ionic strength the activity coefficients are appreciably less than unity. Thus, strictly, as

$$K_S = a_{A^+} a_{B^-} = [A^+][B^-] f_{A^+} f_{B^-}$$

K_S is a constant, and $f_{A^+} f_{B^-} < 1$, $[A^+]$ and $[B^-]$ must be greater than when $f = 1$, i.e. when there is little or no added electrolyte present. For example, the solubility of barium sulphate at $25°$ is $2\frac{1}{2}$ times greater in $0.04M$ potassium nitrate solution than in water and at $37°$ that of calcium oxalate monohydrate in $0.6M$ sodium chloride solution is five times the solubility in water [21]. The magnitude of the effect increases with increasing charge on the ions of the added salt ions, because the effect on the ionic strength, and hence on $f_{A^+} f_{B^-}$, is larger.

In very concentrated electrolyte solutions, activity coefficients are > 1 so the solubility of precipitates is then less than that in the absence of the added salt.

The Precipitation Process [16, 22]

In order that a precipitate shall form, it is not sufficient merely to exceed the equilibrium solubility; an appreciably greater concentration, the super-saturation limit, must be exceeded. Thus, although solutions containing solute in excess of its solubility are thermodynamically unstable, that is, they are metastable, they do not give rise to a precipitate until this supersaturation concentration exceeds the supersaturation limit and the solution becomes 'labile'. Like solubility, the supersaturation limit can also be expressed in terms of an ionic product

$$Q = [M^{n+}]_{ss} [A^{n-}]_{ss} > K_S = [M^{n+}]_{eq} [A^{n-}]_{eq} \qquad (2.85)$$

where ss signifies the supersaturation limit conditions, eq the equilibrium condition and Q is thus defined as the ionic product at the supersaturation limit. In many instances Q considerably exceeds K_S. For example, Q is about $6K_S$ for silver chloride and at least $200K_S$ for barium sulphate. The supersaturation limit generally increases with temperature.

Precipitation can be considered as a two-stage process: the formation of nuclei, followed by their growth into the precipitate particles that are eventually obtained.

Nucleation

Nuclei are the smallest particles of precipitate capable of spontaneous growth in a medium supersaturated with solute. The formation of these particles is termed nucleation and this must take place before precipitation can occur. In the simplest examples of nucleation, ions are assumed to form clusters of increasing size by a series of reversible reactions:

$$M^{n+} + A^{n-} \rightleftharpoons MA$$

$$MA + A^{n-} \rightleftharpoons MA_2^{n-}$$

$$MA_2^{n-} + M^{n+} \rightleftharpoons M_2A_2$$

$$M_{n-1}A_n^{n-} + M^{n+} \rightleftharpoons M_nA_n \text{ (nucleus; } n \sim 6 \text{ for } BaSO_4 \text{ [23])}$$

These clusters are more likely to dissociate than to add more ions, because the stabilization of the ions in solution by hydration and their counter-ion 'atmospheres' exceeds the stabilizing effect of the crystal lattice, which is small in such small particles. Thus the formation of clusters up to the size of the nucleus is energetically unfavourable and requires an ionic product of the reactants much in excess of the solubility product. The ionic product that must be exceeded is Q, given by the supersaturation limit. Therefore, as Q is not exceeded in unsaturated or metastable solutions, no nucleation, and hence no precipitation, occurs in these solutions.

Nucleation, therefore, occurs only when the ionic product is sufficient for the solution to be 'labile'. Moreover, once Q is exceeded, the rate and extent of nucleation increase with increasing reactant concentrations, as would be expected from simple kinetic considerations. Thus, if Q is exceeded only by a small amount, nucleation is slow and the appearance of a precipitate is delayed. The time that elapses between the mixing of reactants and the appearance of a precipitate is called the induction period. Dilute solutions of alkaline earth metal sulphates and lead chloride are systems which have relatively long induction periods.

Nucleation can be facilitated by the introduction of suitable submicroscopic particles into the solution. A commonly used procedure is to produce minute glass fragments by scratching the walls of the precipitation vessel. A similar effect is achieved by adding traces of certain sparingly soluble species. For example, supersaturated metastable solutions of lead sulphate in ammonium acetate solution may be quite stable, but addition of μg amounts of barium ions results in the formation of minute particles of barium sulphate, on which lead ammonium sulphate precipitates [24, 25]. Similarly, small amounts of lithium ions induce the precipitation of sodium iron(III) periodate [26]. In both examples, appearance of the induced precipitate provides a sensitive test for the inductant.

Crystal Growth

The subsequent spontaneous growth of the nuclei to form the final particles of precipitate soon overtakes nucleation as the predominant process in the reaction. It continues until the solubility product is no longer exceeded. In general, the rate of growth of the particles increases with increasing super-saturation, but an increased growth rate also gives rise to an increased number of imperfections in the resulting crystals. Thus, very large supersaturations result in the rapid precipitation of imperfect and extremely small particles, so that an amorphous, gelatinous precipitate is produced.

Completeness of Precipitation

Optimal conditions for complete precipitation may be established from consideration of the equilibria discussed previously (p. 54). However, these alone may not be sufficient for complete precipitation *in a reasonable time*, because the rates of nucleation or crystal growth may be slow, as in the precipitation of calcium tartrate (p. 193).

Purity of Precipitates

A precipitate often carries other species down with it. If precipitation of the main component had not been brought about, the other species would have remained in solution. This process is called **co-precipitation**. There are three main types.

(1) *Bulk co-precipitation*, in which the contaminant is distributed throughout the bulk of the precipitate crystals. This is sometimes described as the formation of solid solutions or of mixed crystals.
(2) *Surface co-precipitation*, in which the contaminant is bound to the surface of the precipitate. This process, known as adsorption, is important when the precipitate is composed of a large number of small particles and hence has a large surface area, as, for example, in a gelatinous precipitate of hydrated iron(III) oxide.
(3) *Inclusion*, in which the solution is mechanically trapped in the growing crystal. This is of greatest importance in the crystallization of relatively soluble salts.

Excessive co-precipitation often occurs in qualitative analysis because precipitation is too rapid and takes place in the presence of contaminants at rather high concentrations. Confirmatory tests for a given ion should be designed, therefore, so that co-precipitatated ions will not interfere. Another deleterious effect of co-precipitation in a systematic scheme of qualitative analysis is the unwanted reduction in concentration of various species as the analysis proceeds, although this is only serious if (a) only traces are present to start with or (b) co-precipitation is almost quantitative.

In quantitative analysis it is usually essential that co-precipitation is mini-mized. This is normally achieved by providing conditions for slow precipitation so that large crystals with a small total surface area are formed. As slow precipi-tation is a consequence of low supersaturation, this can be achieved not only by the use of suitably dilute solutions, but also very effectively by **precipitation from homogeneous solution** [27]. In this technique the conditions are adjusted so that the supersaturation slowly builds up, so that once the value required for nucleation to occur has been reached, crystal growth predominates over further nucleation and well-formed, relatively uncontaminated crystals are often produced. Typical examples are slow production of the precipitant by hydrolysis, e.g. of sulphamic acid to sulphate ions for the precipitation of barium ions [27]:

$$H_2NSO_3^- + H_2O \longrightarrow NH_4^+ + SO_4^{2-} ,$$

the oximation of biacetyl in the presence of nickel ions to precipitate bis-(dimethylglyoximato)nickel [28]:

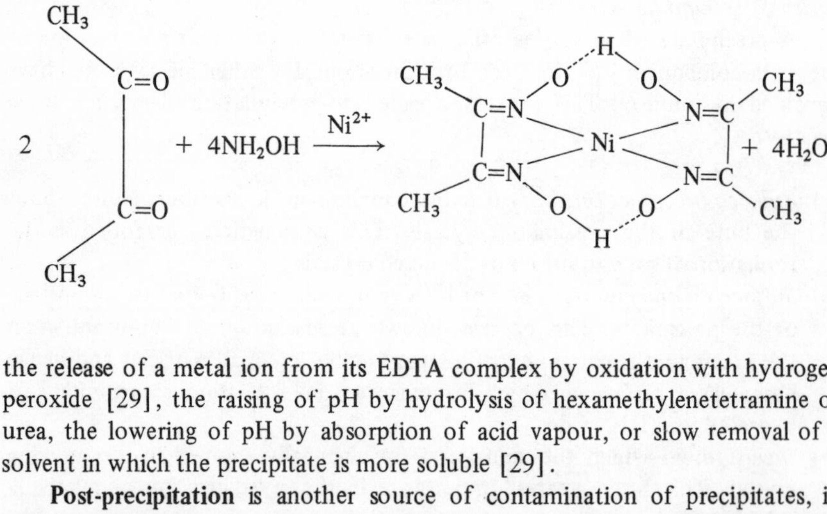

the release of a metal ion from its EDTA complex by oxidation with hydrogen peroxide [29], the raising of pH by hydrolysis of hexamethylenetetramine or urea, the lowering of pH by absorption of acid vapour, or slow removal of a solvent in which the precipitate is more soluble [29].

Post-precipitation is another source of contamination of precipitates, in which a contaminant is deposited onto a precipitate *after* precipitation of the required substance is complete. In some instances, post-precipitation starts immediately after the required precipitation, as, for example, of zinc sulphide after the precipitation of mercury(II) sulphide when zinc is present with mercury(II) in $0.4M$ sulphuric acid. In other instances there is an appreciable delay as, for example, in the post-precipitation of magnesium oxalate on calcium oxalate and of zinc sulphide on bismuth sulphide in $0.2M$ hydrochloric acid.

In qualitative analysis, therefore, precipitates should be separated from their mother liquor as soon as possible, to minimize post-precipitation.

Aging of Precipitates
    Freshly formed precipitates, especially if they consist of small or irregular particles, sometimes change their shape, size, crystal structure or even their

chemical composition. These processes are often accompanied by a marked decrease in solubility. For example, nickel and cobalt sulphides precipitate only in alkaline solutions, but they do not redissolve in acidic solutions. This property has been correlated with the observation that nickel sulphide, when freshly precipitated, is amorphous, but even after five minutes in dilute hydrochloric acid it is appreciably crystalline. Cadmium sulphide can be precipitated from hydrochloric acid as mainly amorphous material. Aging by standing in hydrochloric acid or sulphuric acid markedly increases the crystallinity of the precipitate, but the crystals in hydrochloric acid are mainly hexagonal, whereas those in sulphuric acid are almost entirely cubic [30]. Fortunately, the solubility of both forms is almost the same.

These changes arise because very small particles (those of diameters of a few μm or less) have a greater solubility then macroscopic particles. The mathematical relationship between the solubilities and the particle radius is given by

$$\ln \frac{S}{S_r} \propto r$$

where S_r is the solubility of a particle of radius r, and S is the normal solubility (that of macroscopic crystals). Thus, when a solution is saturated with respect to the larger crystals, it is not saturated with respect to the smaller ones, so the smaller ones dissolve. The solubility of the larger crystals is thus exceeded, and the dissolved material deposits on the larger crystals. This growth of larger particles at the expense of the smaller ones is called **Ostwald ripening**. It occurs, for example, with barium sulphate. It is slow, but can be accelerated by heating. In a similar way, protrusions on an irregular crystal have a solubility higher than that of a flat surface, so that they slowly dissolve and redeposit in depressions and indentations in the crystal, thus producing a more regular crystal. This process is called **internal ripening**. Both processes involve slow dissolution and reprecipitation, so impurities within the crystals can be released and some change of crystal form can occur. Moreover, the resulting reduction in total surface area with formation of larger and more regular crystals reduces the amount of adsorbed impurities.

Sometimes a metastable crystal form is first produced, which is converted into a stable form on aging. The resultant reorganization often leads to a purer precipitate.

Colloidal suspension

A colloidal suspension consists of extremely small particles ($10^{-3} - 10^{-1} \mu$m in diameter) dispersed in the medium. It may be formed as a result of a precipitation or attempted precipitation reaction, or of washing a precipitate with water (peptization). The particles scatter light, so the suspension is 'cloudy' or translucent and cannot be clarified either by filtering or by normal centrifuging.

Very dilute colloidal suspensions may appear to be clear solutions but can be identified by shining a beam of light through them and observing that the light is scattered (best seen at right angles to the beam). This is known as the Tyndall effect.

There are two types of colloid. *Lyophilic* colloids are generally viscous and are insensitive to coagulation by electrolytes. They are highly hydrated, and examples include solutions of natural polymers such as polysaccharides and proteins, and freshly prepared hydrated silica.

Lyophobic colloids are usually formed from inorganic species, and owe their stability to their high electrical charges resulting from the preferential adsorption of either positively or negatively charged ions. In a particular system, therefore, each colloid particle carries a surface charge of the same sign, which is balanced by a diffuse, surrounding layer of oppositely charged 'counter-ions'. These counter-ions, in turn, have their own counter-ion atmospheres. Thus the high electrical potential arising from the charge on the surface of the colloidal particle decreases exponentially with distance from the surface. Provided, therefore, that this potential is sufficient to repel a similar particle at a distance where van der Waals attraction begins to become appreciable, the colloidal suspension is stable.

Lyophobic colloids are much less viscous than lyophilic colloids of the same concentration. They can be flocculated (that is precipitated) by adding electrolytes. The ions in the electrolyte responsible for flocculation are those of opposite charge to the charge on the colloidal particles. An increase in the concentration of electrolyte compresses the diffuse layer around the particles, thus allowing the particles to approach to a distance where the van der Waals attraction exceeds the electrostatic repulsion. Multicharged ions are especially effective in compressing the diffuse layer, so the concentration of an ion required to cause flocculation decreases markedly with increasing ionic charge. The relative molar concentrations required are approximately [31]: M^+, 1.0; M^{2+}, 0.016; M^{3+}, 0.0013; M^{4+}, 0.00024. Heating and shaking, which facilitate the close approach of colloid particles, also aid flocculation.

Flocculation can often be at least partly reversed to reform the colloidal suspension by washing the precipitate with pure water. This process is called **peptization**. The water dilutes the coagulating electrolyte so that the particles repel one another again and become dispersed. If peptization is likely, washing should be done with a suitable electrolyte solution. However, it is often difficult to remove multivalent cations in this way, and, in addition, aging processes may have occurred that prevent peptization from taking place.

2.3.4 Redox Reactions

Redox reactions involve the transfer of electrons between reactants (p. 32), and in common with other types of reaction can be interpreted quantitatively. Many redox reactions, however, involve a sequence of steps, some of which may be too slow for equilibrium conditions to be established in the time available.

Equilibrium data, therefore, should be used for calculations only when it is certain that equilibrium exists, and the stoichiometry of the overall reaction is known.

A redox reaction between iron(II) and cerium(IV) ions may be formally expressed as:

$$Fe^{2+} + Ce^{4+} \rightleftharpoons Fe^{3+} + Ce^{3+}$$

although the actual chemical form of an interacting species is likely to be different [e.g. $Ce(SO_4)_3^{2-}$].

The equilibrium constant for this reaction is:

$$K = \frac{[Fe^{3+}][Ce^{3+}]}{[Fe^{2+}][Ce^{4+}]} \tag{2.86}$$

The reaction may be considered as two half-reactions:

$$Fe^{3+} + e \rightleftharpoons Fe^{2+}$$

$$Ce^{4+} + e \rightleftharpoons Ce^{3+}$$

in which each half-reaction involves a change in oxidation state. The tendency for either half-reaction to take place can be measured in terms of its **electrode potential**, which is independent of the other accompanying half-reaction. Any half-reaction in aqueous solution can be written in the general form as:

$$Ox_{aq} + ne \rightleftharpoons Red_{aq} \tag{2.87}$$

where Ox is the oxidized species and Red the reduced form. The electrode potential of a half-reaction is the potential of an inert metal electrode, such as platinum, when immersed in a solution containing both the oxidized and reduced species. The potential arises because of the ability of a reductant or an oxidant to transfer electrons to or from (respectively) the electrode. The greater the oxidizing power, for example, the greater the tendency to remove electrons from the metal, and the greater the potential to which the electrode is raised. This potential can be calculated from the Nernst equation:

$$E = E^\circ + \frac{RT}{nF} \ln \frac{a_{Ox}}{a_{Red}} \tag{2.88}$$

where n is the number of electrons transferred in the half-reaction, R is the gas constant, T the absolute temperature and F the Faraday. E°, the **standard electrode potential**, is defined as the electrode potential when the reduced and oxidized forms in solution are at unit activity.

Single electrode potentials cannot be determined because only the potential difference between two electrodes connected to complete an electrochemical cell can be measured.

For purposes of assigning individual electrode potentials a standard reference electrode is chosen and defined as having zero potential. This electrode is the standard hydrogen electrode, although in practice it is more convenient to use a saturated calomel (Hg_2Cl_2) electrode, the potential of which is accurately known by measurement against a standard hydrogen electrode.

Several conventions are used for describing electrodes and electrode reactions, and particularly the sign of the electrode potential. The convention proposed by the International Union of Pure and Applied Chemistry (IUPAC) will be used here [32]. In this convention the oxidized state is written on the left-hand side of the equation, as in Eq. (2.87). E° is the potential difference for the cell:

hydrogen electrode

$$Pt \left| \begin{matrix} Ox_{aq} \ (a=1) \\ \\ Red_{aq} \ (a=1) \end{matrix} \right\| \begin{matrix} H^+_{aq} \\ \\ (a=1) \end{matrix} \left| \begin{matrix} Pt \\ \\ H_2 \ (1 \ atmos.) \end{matrix} \right.$$

and has the same sign as the electrode in the left-hand cell (i.e. it is negative if this electrode is the negative pole of the cell). The single vertical lines represent the platinum/solution interfaces and the double vertical line the liquid junction between the two half-cells.

An electrochemical cell, like a redox reaction in a single solution, involves two half-reactions, and equilibrium is reached when each half-reaction has the same tendency to occur; that is, when each half-reaction has the same electrode potential and the cell has a net potential of zero. When reactions proceed in electrochemical cells the transfer of electrons from the reduced to the oxidized species takes place through the electrodes, which are connected by an outside circuit, producing electrical work. The same reactants may be mixed to form a homogeneous solution. The energy produced by direct electron transfer between the reactants in solution is released as an equivalent amount of thermal energy. The energy expended and the equilibrium concentrations are the same in both instances.

Unlike other types of reaction, it is unnecessary to measure the equilibrium constant for each overall redox reaction because data for the appropriate half-reactions can be combined to calculate the equilibrium concentrations. Electrode potentials are thus a convenient, readily tabulated measure of equilibrium constants for half-reactions. This may be illustrated by calculating the equilibrium constant for the cerium(IV)/iron(II) reaction. As the solutions are moderately dilute, concentrations are conveniently used in place of activities in calculations.

At equilibrium, if the reaction between cerium(IV) and iron(II) is carried out in an electrochemical cell, the electrode potentials of the two half-reactions are equal. Thus

$$E = E^{\circ}_{(Ce^{4+}/Ce^{3+})} + \frac{RT}{F} \ln \frac{[Ce^{4+}]}{[Ce^{3+}]} = E^{\circ}_{(Fe^{3+}/Fe^{2+})} + \frac{RT}{F} \ln \frac{[Fe^{3+}]}{[Fe^{2+}]}$$

$$E^{\circ}_{(Ce^{4+}/Ce^{3+})} - E^{\circ}_{(Fe^{3+}/Fe^{2+})} = \frac{RT}{F} \ln \frac{[Fe^{3+}][Ce^{3+}]}{[Fe^{2+}][Ce^{4+}]}$$

$$= 2.3 \frac{RT}{F} \log_{10} \frac{[Fe^{3+}][Ce^{3+}]}{[Fe^{2+}][Ce^{4+}]}$$

$$= 0.059 \log K \text{ [from Eq. (2.86)] at } 25^{\circ}$$

(at 25° the constant $2.3\,RT/F$ has a value of 0.059 V).

From tables of standard electrode potentials [33], $E^{\circ}_{(Ce^{4+}/Ce^{3+})} = 1.45$ V and $E^{\circ}_{(Fe^{3+}/Fe^{2+})} = 0.76$ V. Hence 0.059 log $K = 1.45 - 0.76 = 0.69$ and log $K = 11.7$. Thus $K = 10^{11.7} = 5.0 \times 10^{11}$. This very large value for the equilibrium constant implies that iron(II) is virtually completely oxidized by an equal concentration of cerium(IV). This is the basis of the titrimetric determination of iron(II) with cerium(IV).

The addition of an excess of a reactant affects the equilibrium concentrations and the amounts of the other species present which are consumed or produced as the reaction proceeds to equilibrium. For example, suppose a solution is initially $10^{-3}M$ in both iron(II) and cerium(IV).

When equilibrium is achieved,

$$[Fe^{2+}] = [Ce^{4+}] \text{ and } [Fe^{3+}] = [Ce^{3+}]$$

As $K = 10^{11.7}$, the iron(II) and cerium(IV) concentrations at equilibrium are very small, so the approximation can be made that $[Fe^{3+}] = 10^{-3}M$. Thus, from Eq. (2.86)

$$K = \frac{[Fe^{3+}]^2}{[Fe^{2+}]^2} = \frac{(10^{-3})^2}{[Fe^{2+}]^2} = 10^{11.7}$$

Therefore, $[Fe^{2+}]^2 = 10^{-17.7}$ and $[Fe^{2+}] = [Ce^{4+}] = 10^{-8.9}M = 1.3 \times 10^{-9}M$.

However, if the solution is initially $10^{-3}M$ in iron(II) and $0.1M$ in cerium(IV), only 1% of the cerium(IV) is consumed in the reaction, so that, at equilibrium,

$$[Ce^{4+}] = 0.99 \times 10^{-1}M \sim 0.1M$$

$$[Ce^{3+}] = [Fe^{3+}] = 10^{-3}M$$

Therefore, from Eq. (2.86)

$$K = \frac{(10^{-3})^2}{[Fe^{2+}] \times 0.1} = 10^{11.7}$$

and $$[Fe^{2+}] = \frac{10^{-6}}{10^{11.7} \times 10^{-1}} = 10^{-16.7}M = 2 \times 10^{-17}M \ .$$

This concentration should be contrasted with the value obtained above for the stoichiometric equilibrium mixture ($1.3 \times 10^{-9}M$).

The electrode potentials of redox couples (such as Fe^{3+}/Fe^{2+} or MnO_4^-/Mn^{2+}) are markedly influenced by a variety of factors. For example, changes in hydrogen-ion concentration may not only affect the electrode potential, but change the stoichiometry of the reaction. The oxidation potential of the permanganate ion [manganese(VII)], for example, is high in acidic solutions, in which reduction is to manganese(II). As the pH is increased, the oxidation potential decreases, and at pH ca. 7 reduction stops at manganese(IV) or, at high pH (ca. 13), at manganese(VI) (manganate ions). Thus the number of electrons which can be accepted by each permanganate ion changes from five in acidic media to one in strongly alkaline media.

The chemical entities involved in redox reactions are often present as complex ions, and the electrode potential will depend on the relative strengths of the complexes formed between the ligands and the reduced and oxidized species. For instance, the complexes formed between fluoride ions and iron(III) ions (FeF^{2+}, FeF_2^+, etc.) are appreciably stronger than those formed with iron(II) ions. Thus the concentration of uncomplexed iron(III) ions is much decreased by the complexing action of the fluoride ions, unlike that of iron(II) ions. Thus $[Fe^{3+}]$ is decreased and so is the potential for the iron(III)/iron(II) couple [Eq. (2.88)]. Consequently the equilibrium constant in (for example) the cerium(IV)-iron(II) reaction is increased.

Formal potential

Calculations using electrode potentials are restricted to systems in which the concentrations of discrete ionic species are known or may be calculated. In many systems, however, the oxidant or reductant may exist in a number of different forms in solution, especially in the presence of complexing agents, and the concentration of each form may not readily be calculated. Moreover, in solutions of high ionic strength the use of concentrations in place of activities in calculation results in considerable error. To facilitate the use of electrode potentials under such circumstances, the electrode potentials for many redox systems have been measured at various ionic strengths and pH values, and in the presence of various complexing agents. The value obtained in each medium is called the **formal electrode potential** or **conditional electrode potential** for the particular redox couple in that medium. For example, the standard electrode potential for the Fe^{3+}/Fe^{2+} couple is 0.76 V. The formal potential for the couple in $0.1M$ hydrochloric acid is 0.73 V, in $1M$ hydrochloric acid 0.70 V, and in 0.1-$4M$ sulphuric acid 0.68 V. The formal potential for the Ce^{4+}/Ce^{3+} system in $1M$ sulphuric acid is 1.44 V. These values of the formal potential can be used in the Nernst equation for the calculation of the equilibrium potentials [Eq. (2.88)]. By substitution of these values in the equation derived on p. 67, the equilibrium constant for the reaction in $1M$ sulphuric acid can be calculated:

$$0.059 \log K = 1.44 - 0.68 = 0.76$$

$$\log K = \frac{0.76}{0.059}$$

$$K = 7.6 \times 10^{12}$$

This value should be compared with that previously obtained by using standard electrode potentials: 5.0×10^{11} (p. 67).

Such conditional electrode potentials are readily evaluated and handled by means of the α-coefficient method (p. 56). Consider the Fe^{3+}/Fe^{2+} system in chloride medium, assuming that only the $FeCl_4^-$ complex is significant. The Nernst equation gives

$$E = E^{\circ}{}_{Fe^{3+}/Fe^{2+}} + 0.059 \log \frac{[Fe^{3+}]}{[Fe^{2+}]} \qquad (2.89)$$

In presence of chloride, however, the iron(III) is present as more than one species. For simplicity let us suppose these are Fe^{3+} and $FeCl_4^-$. Then

$$C_{Fe^{3+}} = [Fe^{3+}] + [FeCl_4^-]$$

$$= [Fe^{3+}] \{1 + \beta_4 [Cl^-]^4\}$$

$$= [Fe^{3+}] \, \alpha_{Fe(Cl)}$$

where β_4 is the overall stability constant for $FeCl_4^-$. Hence Eq. (2.89) becomes

$$E = E^{\circ}{}_{Fe^{3+}/Fe^{2+}} + 0.059 \log \frac{C_{Fe^{3+}}}{[Fe^{2+}] \alpha_{Fe(Cl)}}$$

and we can split the second term on the right-hand side into two:

$$E = E^{\circ}{}_{Fe^{3+}/Fe^{2+}} + 0.059 \log \frac{1}{\alpha_{Fe(Cl)}} + 0.059 \log \frac{C_{Fe^{3+}}}{[Fe^{2+}]} \qquad (2.90)$$

and then combine the first two terms to obtain a conditional standard potential $E^{o\prime}$ which is clearly a function of the chloride ion concentration:

$$E = E^{o\prime}_{Fe^{3+}/Fe^{2+}} + 0.059 \log \frac{C_{Fe^{3+}}}{[Fe^{2+}]} \qquad (2.91)$$

The value of $E^{o\prime}$ decreases as $[Cl^-]$ is increased.

A similar argument can be used to show the effect of precipitation or of protonation.

2.3.5 Partition Equilibria and Liquid–Liquid Extraction [34-37]

In many chemical reactions in water a reactant or reaction product may be readily soluble in a certain organic solvent, and may be extracted from the aqueous medium by shaking with the organic solvent if this is not miscible with water. For example, the complex formed by iron(III) ions in strongly acidic solutions of chloride ions, and the zirconium-Alizarin Red S complex formed in $1M$ perchloric acid, are both soluble in water, but are also soluble in ethers and in amyl alcohol, respectively. Furthermore, in aqueous solution the reactions:

$$H^+ + C_6H_5COO^- \rightleftharpoons C_6H_5COOH(s)$$

and
$$Al^{3+} + 3C_9H_6NO^- \rightleftharpoons Al(C_9H_6NO)_3(s)$$
$$\text{8-hydroxy-}$$
$$\text{quinolinate}$$

both produce precipitates which are readily soluble in chloroform or benzene. In all these examples the organic solvent is only slightly miscible with water. Thus, if the aqueous solution containing the soluble or precipitated species is intimately shaken with the organic solvent, this species will dissolve in the organic phase. This process is known as **liquid-liquid extraction** (or solvent extraction).

As with other chemical reactions, the relative concentrations of a particular species A in each phase are governed by an equilibrium:

$$A_{aq} \rightleftharpoons A_{org}$$

where the subscripts refer to the aqueous and organic solutions. The equilibrium constant for this reaction:

$$P = \frac{[A_{org}]}{[A_{aq}]} \qquad (2.92)$$

is known as the **partition coefficient**. If A is a simple solvated species that does not dissociate or associate in either the aqueous or the organic phase, the partition coefficient gives a reasonably complete description of the equilibrium. Moreover, if both phases are saturated, as, for example, when solid A is also present,

$$P = \frac{S_{org}}{S_{aq}}$$

(2.93)

where S is the solubility of A in the solvent indicated by subscript.

However, in most analytically useful examples of solvent extraction, the solute is not present as one species, but as molecules or complex ions and their reaction products which have been formed by reversible reactions in the aqueous (and sometimes the organic) phase. For example, the extraction of benzoic acid is affected by its dissociation in aqueous solution; the amount extracted depends on the concentration of undissociated benzoic acid present and hence on the pH of the aqueous solution. Such effects are most easily dealt with by defining a new quantity, the **distribution coefficient** D. The distribution coefficient is defined as the ratio of the total concentration of all species of A in the two phases:

$$D = \frac{\Sigma [A]_{org}}{\Sigma [A]_{aq}}$$

(2.94)

If the benzoic acid is present as the undissociated acid in the organic phase and as a mixture of the undissociated and dissociated forms in the aqueous phase, then

$$D = \frac{[C_6H_5COOH]_{org}}{[C_6H_5COOH]_{aq} + [C_6H_5COO^-]_{aq}}$$

(2.95)

As the dissociation constant is

$$K_a = \frac{[C_6H_5COO^-][H^+]}{[C_6H_5COOH]}$$

it is readily seen that

$$D = \frac{[C_6H_5COOH]_{org}}{[C_6H_5COOH]_{aq}\{1 + K_a/[H^+]\}} = \frac{P}{1 + K_a/[H^+]}$$

(2.96)

and the extraction becomes less efficient as the pH is increased. Similar calculations can be carried out for the partition coefficient of $Al(C_9H_6NO)_3$ under the influence of stepwise complexing equilibria, ligand dissociation, and partition of the ligand itself between the two solvents. For tris(8-hydroxyquinolinato) aluminium, assuming the uncharged chelate to be the only extractable aluminium-containing species,

$$D = \frac{\Sigma [Al]_{org}}{\Sigma [Al]_{aq}}$$

$$= \frac{[Al(C_9H_6NO)_3]_{org}}{[Al(C_9H_6NO)_3]_{aq} + [Al(C_9H_6NO)_2^+]_{aq} + [Al(C_9H_6NO)^{2+}]_{aq} + [Al^{3+}]_{aq}}$$

$$(2.97)$$

In this instance, aluminium is the species of analytical interest, so the denominator is the total concentration of aluminium in the aqueous phase. Distribution coefficients vary with acidity, ligand concentration, and other parameters that affect the formation of the extractable species. These effects can be incorporated in Eq. (2.97) by utilizing the equilibria governing the formation of the various complexes and the protonation of the ligand, so that their effect on the distribution coefficient, D, can be calculated.

For successful selective extraction of one compound or metal-ion complex from a mixture it is necessary for the distribution coefficient for that compound or metal-ion complex to be large ($\gg 1$) and those of the other compounds or metal-ion complexes also present in the mixture to be small ($\ll 1$). Such conditions are dependent not only on the partition coefficients but also on the values of the dissociation constants where acids and bases are involved, and hence on the pH of the solution. The pH is also important in the selective extraction of metal-ion complexes because it often influences the degree of formation of the extractable complex (see also p. 51) and thus the amount of complex available for extraction into the organic phase. For example, the effect of pH on the extraction of metal 8-hydroxyquinolinates is shown in Fig. 2.4. Generally, with these complexes, the more stable the chelate, the lower is the pH at which extraction is possible and at which it is complete. This is a direct consequence of the fact that the degree of chelate formation at low pH values is larger for the stronger complexes than for the weaker. From Fig. 2.4 it can be seen that at pH 2.0, copper is almost 100% extracted, whereas zinc, cadmium, lead and silver are not extracted, so copper may thus be separated from these ions under such conditions.

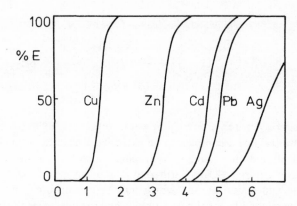

Fig. 2.4 – Effect of pH on the extraction of Cu^{2+}, Ag^+, Zn^{2+}, Cd^{2+} and Pb^{2+} by 0.10M 8-hydroxyquinoline in chloroform.

Nature of the Extraction Process [34]

Ions and molecules dissolved in aqueous solution are usually associated with a sheath of water molecules; the close compatibility between these hydrated species and the water molecules confers stability on the separate ions or molecules of solute so that they do not tend to aggregate and form a separate phase, by forming a precipitate, for example. Water is particularly effective in this respect because its high dielectric constant reduces the coulombic interaction between ions and between dipolar molecules. Similar solvents, but with lower dielectric constants, such as ethanol, are much less effective in promoting charge separation, so that ionic species are much less soluble in ethanol than in water. The solubility of ionic species in organic solvents of very low dielectric constant, such as benzene or chloroform, is even smaller. The dissolution of organic compounds in suitable organic solvents is commonplace, and can usually be predicted by the principle 'like dissolves like' [38]; for example, benzoic acid is soluble in benzene. If such a compound is also soluble in water, it will partition between water and the non-aqueous solvent. On transfer of the species from water to the organic phase, some of the associated water molecules will be displaced or removed from the hydration layer of the compound, making the compound more compatible with the organic phase.

To enable ionic compounds to dissolve in organic solvents of low dielectric constant, it is usually necessary to form either an uncharged complex or an ion-pair which may dissolve without charge separation. For the first purpose an organic ligand is usually employed, which displaces co-ordinated water molecules from the ion and counterbalances the ionic charge. Such systems are almost entirely restricted to metal ions. For example, tris(8-hydroxyquinolinato)-aluminium is formed by displacement of six water molecules from the co-ordination sphere of the aluminium ion by three 8-hydroxyquinolinate ions to form the uncharged tris-chelate:

$$Al(H_2O)_6^{3+} + 3 \quad \rightleftharpoons \quad + 6H_2O$$

This complex is almost entirely insoluble in water, but is readily extracted into many organic solvents. For a complex to be extracted it should be uncharged, hence the ligand should not contain hydrophilic ionic groups such as the charged sulphonate group ($-SO_3^-$). Furthermore, dipolar groups such as the hydroxy, carbonyl and amino groups, which may be present but not involved in binding with the metal ion, are also hydrophilic and may decrease the extractability of the complex from aqueous solution. If an uncharged complex forms without displacement of all the water molecules co-ordinated to the metal ion, the complex will be more readily extractable into polar organic solvents, such as alcohols and ketones, than into non-polar solvents such as benzene and carbon tetrachloride. This arises because the polar water molecules are displaced more readily by polar then by non-polar organic molecules, which enhances the solubility of the complex in polar organic solvents. For example, the chelate of zinc with 8-hydroxyquinoline:

$$Zn(H_2O)_6^{2+} + 2C_9H_6NO^- \quad \rightleftharpoons \quad + 4H_2O$$

retains two molecules of water in the complex, which is thus more soluble in methyl ethyl ketone than in chloroform. An alternative is to displace the water molecules by forming a neutral adduct with the reagent itself or with an organic molecule that is less polar than water, for example $Sr(C_9H_6NO)_2.(C_9H_7NO)$ or $Sr(C_9H_6NO)_2.(C_4H_9NH_2)_2$.

The solubility of ions as ion-pairs in organic solvents of low dielectric constant can be increased in the following ways.

(1) *By the use of bulky, organic counter-ions.* These form ion-pairs which dissolve without charge separation because of the essentially organic nature of the large counter-ion. For example, tetraphenylphosphonium or arsonium

cations are used for the extraction of certain anions, and tetraphenylborate anions for some cation extractions. Large dye cations can be used in a similar way for extraction of anions. Details may be found in two review articles [39, 40].

(2) *By the use of organic solvents to solvate ions.* For example, iron(III) in fairly concentrated hydrochloric acid forms the tetrachloroferrate(III) anion, $FeCl_4^-$, which may be extracted into an ether or suitable ester as its association compound with solvated hydrogen ions.

With ethers (R_2O), the anion may be extracted as an etherate [41], of possible structure:

$$[R_2OH^+.Fe(R_2O)_2Cl_4^-] \quad \text{or} \quad [H_3O^+.Fe(R_2O)_2Cl_4^-]$$

(3) *By the use of liquid ion-exchangers* [36, 42]. Long-chain organic amines are soluble in organic solvents, but not in aqueous media. However, when such compounds are protonated, the resulting ions become water-soluble, but ion-pairs formed with suitable anions are also extractable into organic solvents. Such anions include anionic metal complexes such as $AuCl_4^-$. The extracted counter-anion can be displaced by another anion which forms a more stable ion-pair with the protonated organic molecule. Hence such organic compounds are referred to as liquid ion-exchangers. Quaternary ammonium salts function in the same way.

(4) *By salting-out effects.* The addition of electrolytes in sufficient amounts to give high concentrations usually enhances the extraction of ion-association complexes. This enhancement is termed a **salting-out effect**. For example, the extraction of uranyl or thorium nitrate from acidic solution is greatly enhanced by the addition of large amounts of metal nitrates. Salts of multicharged cations are particularly effective. The functions of the salting-out agent include (a) providing a large excess of the complexing anion, thus increasing the concentration of the complex, (b) the binding of water molecules, thus making them unavailable for hydration of the complex, and (c) reducing the dielectric constant, thus favouring ion-association complex formation.

Solvent Extraction in Qualitative Analysis

Solvent extraction can be used instead of precipitation as the separational tool in qualitative analysis [43, 44].

2.4 MULTIPLE EQUILIBRIA

In the preceding discussion many types of equilibria have been described and examples have been given of the influence of a combination of some of these equilibria on analytical reactions. These include the effect of acidity on

complex (p. 51) and precipitate (p. 56) formation, the effect of complex formation on redox equilibria (p. 68), and the effect of acid-base and complexation
equilibria on solvent extraction (p. 72). Consideration of these factors emphasizes
that when an equilibrium calculation is attempted for any system, all the
equilibria influencing the analytical reaction must be taken into account if
meaningful answers are to be achieved. In such instances, a complete mathematical evaluation of the equilibria may require extensive computation, and
the use of an appropriate computer program [11] will be beneficial.

Many examples of calculations involving multiple equilibria will be found
in Chapter 3.

2.4.1 Masking and Demasking

Particularly useful applications of multiple equilibria are the processes of
masking and demasking [21, 45]. Metal ions can often be prevented from
participating in a particular chemical reaction by complexing the ion with a
suitable ligand. For example, silver chloride does not precipitate when the
silver ion is complexed by ammonia. In such circumstances, the silver ion is said
to be **masked**.

Masking occurs when the complex formed is sufficiently stable to resist
breakdown in the presence of a competing reagent. For example, the following
equilibria are involved in a solution containing silver and chloride ions and
ammonia:

$$Ag^+ + 2NH_3 \rightleftharpoons [Ag(NH_3)_2]^+$$

$$Ag^+ + Cl^- \rightleftharpoons AgCl(s)$$

If the masking of the silver ions by ammonia is to be effective in preventing the
precipitation of silver chloride, the free silver-ion concentration, when multi-
plied by the particular chloride-ion concentration, must be insufficient to exceed
the solubility product of silver chloride. For the silver ammine solution discussed
on p. 50 ($10^{-4}M$ ammine, $10^{-2}M$ ammonia), the silver ion concentration was
calculated to be $5 \times 10^{-8}M$. As the solubility product, K_S, for silver chloride is
2×10^{-10} at 25°, precipitation will not occur in such a solution until (neglecting
supersaturation effects):

$$[Cl^-] \geqslant \frac{K_S}{[Ag^+]} = \frac{2 \times 10^{-10}}{5 \times 10^{-8}} = 4 \times 10^{-3}M$$

Thus $10^{-2}M$ ammonia masks a $10^{-4}M$ silver ion solution against precipitation by
a chloride solution that is $< 4 \times 10^{-3}M$. Note, however, that masking is not
complete if the chloride ion concentration is $\geqslant 4 \times 10^{-3}M$, and it is possible to

have a precipitate of silver chloride in the presence of free ammonia. Tests based on solubility of silver chloride in ammonia solution depend on a sufficient excess of ammonia being added.

Ammonia is much less effective in preventing silver bromide precipitation. K_S for silver bromide is 8×10^{-13} at $25°$, so the bromide ion concentration required to precipitate silver bromide from the ammine solution considered above is much less than the chloride ion concentration required $(4 \times 10^{-3}M)$:

$$[Br^-] \geqslant \frac{K_S}{[Ag^+]} = \frac{8 \times 10^{-13}}{5 \times 10^{-8}} = 1.6 \times 10^{-5}M$$

Masking is often used to prevent metal ions from interfering in tests. For example, silver can be detected by the red-violet colour it gives with p-dimethyl-aminobenzylidenerhodanine in $2M$ nitric acid. Mercury(II) gives a similar colour but not in the presence of cyanide ions [46] because it is masked as undissociated mercury(II) cyanide, $Hg(CN)_2$. Iodate and periodate ions react with iodide ions in dilute acidic solutions to liberate iodine:

$$IO_3^- + 5I^- + 6H^+ \rightleftharpoons 3I_2 + 3H_2O$$

$$IO_4^- + 7I^- + 8H^+ \rightleftharpoons 4I_2 + 4H_2O$$

but in the presence of molybdate ions, periodate ions form a heteropoly anion, 6-molybdoperiodate [47], which does not react with iodide ions, thus allowing iodate ions to be detected by reaction with iodide ions, in the presence of periodate ions [48]:

$$8H^+ + IO_4^- + 6MoO_4^{2-} \rightleftharpoons I(MoO_4)_6^{5-} + 4H_2O$$

Although masking can be of great value in qualitative or quantitative analysis, undesirable masking effects may occur if the test solution contains complexing agents, which may jeopardize the analysis. Some of the so-called 'interfering anions' in systematic qualitative analysis complex with certain metal ions and prevent them from undergoing the desired reactions. For example, fluoride, citrate and tartrate ions form strong complexes with iron(III), aluminium and chromium(III) ions, so that these metals are not precipitated on addition of hydroxide ions. In such instances, the metals must be **demasked**, that is, removed from their complexes, before they can be precipitated. There are several ways in which this can be done.

(a) By destroying the ligand. Repeated evaporation with concentrated nitric acid destroys organic anions.

(b) By removing the ligand from solution. Fluoride ions can be removed as hydrogen fluoride by volatilization from concentrated hydrochloric acid medium.

(c) By chemical conversion into a non-masking species. Cyanide ions are converted by formaldehyde into cyanohydrin, so that zinc, for example, is released from its cyanide complex:

$$H^+ + CN^- + HCHO \longrightarrow HOCH_2CN$$

(d) By adding a metal ion that complexes more strongly with the ligand. Nickel or magnesium ions displace barium ions from their EDTA complex so that, if sulphate ions are also present, barium sulphate precipitates [49].

There are also more subtle forms of masking. Most complexes are **labile**, that is, they are rapidly formed or destroyed, whereas others are **inert**, and are formed or decomposed very slowly (p. 52), even when they are thermodynamically unstable. It is thus possible to use **kinetic masking** by forming an inert complex. A variant is to use **temperature control**, since reaction rates are reduced by lowering the temperature. A particularly insidious and often unsuspected form of masking is formation of inert heteronuclear ligand-bridged complexes, such as the aluminium–tartrate–uranyl or copper(I)–citrate–chromium(III) complexes. The difficulty is that any excess of one of the two metal species will remain free and detectable, while the other partner will escape detection.

2.5 ORGANIC REAGENTS [50-53]

A wide variety of organic compounds give sensitive and often very selective visual reactions with inorganic and organic species. Such reactions are extensively used for qualitative and quantitative analysis, and numerous examples of their qualitative applications are given in a companion publication [54]. By use of organic reagents, positive identification of most inorganic species is possible without complete separation from other components, so identification procedures may be simplified and speeded up. For example, with *p*-dimethylamino-benzylidenerhodanine, mercury may be detected (red colour) in the presence of bismuth, cadmium and copper. With dimethylglyoxime, nickel (red precipitate) may be detected in the presence of cobalt, zinc and manganese. Because of the high sensitivity of some of these reagents only small amounts are needed. The high sensitivity means that care should be taken in qualitative analysis because species present in only trace amounts may give a positive reaction. It is necessary in such instances to ensure that the major component and the trace component can be distinguished. It may even happen that a trace amount will give the desired reaction, but a large amount gives a different reaction, so that the species is thought to be absent because the expected positive response is not obtained.

Many organic reagents are only sparingly soluble in water, and have to be dissolved in organic solvents, but this rarely leads to significant difficulties. Also, solutions of organic reagents are usually less stable than solutions of inorganic compounds, and decomposition over a period of weeks, possibly by oxidation or hydrolysis, is not uncommon.

Organic reagents have three main uses, as follows.

(1) *For complexing with metal ions*

Such reactions may result in

(a) *precipitation* (precipitants): e.g. aluminium with 8-hydroxyquinoline, nickel with dimethylglyoxime;

(b) *colour formation* (chromogenic reagents): e.g. magnesium with Titan Yellow, cobalt with nitroso-R salt;

(c) *formation of extractable complexes* (extractants): e.g. magnesium with 8-hydroxyquinoline, extracted into chloroform; vanadium(V) with acetylacetone, extracted into benzene;

(d) *masking* (masking agents see p. 76): e.g. tartrate ions mask iron(III); against hydroxide precipitation; EDTA masks barium against sulphate precipitation; EDTA also masks bismuth against colour formation with diethyldithiocarbamate, thus allowing copper to be detected and determined in the presence of bismuth;

(e) *changing redox potentials:* e.g. 1,10-phenanthroline raises E° for the Fe^{3+}/Fe^{2+} couple but lowers that for the Co^{3+}/Co^{2+} couple.

(2) *For non-complexing reactions, mainly with inorganic ions*

Many anions can be precipitated by organic reagents. Sulphate ions, for example, are precipitated by benzidine and its derivatives, such as 4-chloro-4'-ammoniumbiphenyl ions [2]; nitrate ions are precipitated by certain naphthylmethylammonium derivatives [55]. Many organic dyestuffs exist as large cations, and can be used to precipitate or to form extractable ion-pairs with anions [40]. Brilliant Green, a triphenylmethane dyestuff, for example, forms a green compound with the hexachloroantimonate(V) ion, $SbCl_6^-$, which can be extracted

Brilliant Green

into chloroform. Finally, cations such as tetraphenylphosphonium and tetra-phenylarsonium [39] may be used to precipitate anions, such as permanganate and perchlorate, and also anionic metal complexes, such as tetrathiocyanato-cobaltate(II), $Co(SCN)_4^{2-}$, or in extraction of such ions. Large organic anions such as tetraphenylborate can similarly be used to precipitate cations, including some alkali metals.

$$(C_6H_5)_4As^+ + ClO_4^- \longrightarrow (C_6H_5)_4As^+ClO_4^-(s)$$

$$K^+ + (C_6H_5)_4B^- \longrightarrow K^+(C_6H_5)_4B^-(s)$$

(3) For conversion into other compounds

The reaction of hydrazine with salicyaldehyde to form salicylaldehyde hydrazone (p. 34), and the formation of amyl acetate from amyl alcohol (p. 34), are two examples of this type of reaction that have already been noted. Others include the detection of bromine by the conversion of yellow fluorescein into red eosin; of nitrite ions by diazotization and subsequent coupling reactions (Griess reaction), and the detection of sulphur dioxide by the reduction of fuchsin to give a violet product. Details of these reactions, and many others, are given in the companion book [54].

In the subsequent discussion, the use of organic complexing agents is mainly considered.

2.5.1 Organic Complexing Agents

The difference in stability of complexes formed between organic complexing agents and metal ions is an important factor in determining the selectivity of these reactions in qualitative analysis. However, the choice of appropriate reaction conditions to minimize, or exclude, similar unwanted complex forma-tion with other ions, is also important. The desired selectivity is usually attained by adjusting the pH or by masking.

Adjustment of pH

The effect of pH on complex formation, and on precipitation and solvent-extraction equilibria has already been discussed (pp. 51, 56 and 72). A pH is chosen at which the analytical reaction (precipitation, colour formation, extrac-tion, etc.) occurs with the metal ion of interest but with no (or few) other ions. The following two examples have been given previously (pp. 58, 72). Calcium oxalate starts to precipitate at pH 0.5 and its precipitation is complete at pH ~ 2, whereas magnesium does not precipitate below pH 2.8 (p. 58). Copper is extracted as its 8-hydroxyquinolinate into chloroform at pH 2.0 whereas zinc and cadmium are not extracted at this pH (p. 72).

Masking [45]

The reaction of potentially interfering ions with an organic ligand (or other reagent) may be prevented by masking (p. 76). For example, Table 2.3 shows that the solubility of lead oxalate is less than that of calcium oxalate. However, on addition of EDTA and oxalate ions to a solution of calcium and lead ions, only calcium oxalate precipitates if the pH and reagent concentrations have suitable values. For example, consider the concentrations used in Table 2.3 ($[Pb^{2+}] = [Ca^{2+}] = 10^{-3}M$, $[H_2Ox] = 10^{-2}M$). If calcium oxalate is 99.9% precipitated, the residual total oxalate concentration is $9 \times 10^{-3}M$, and the calcium concentration is $10^{-6}M$, so the conditional solubility product is

$$K'_{S_{CaOx}} \leqslant [Ca^{2+}] [Ox^{2-}]$$

$$\leqslant 10^{-6} \times 9 \times 10^{-3}$$

$$\leqslant 9 \times 10^{-9}$$

As $K'_S = K_S \alpha_{Ox(H)}$, $\alpha_{Ox(H)}$ must not exceed $9 \times 10^{-9}/2 \times 10^{-9} = 4.5$, i.e. the pH must be above 3.2. It is simultaneously necessary that for lead oxalate the conditional solubility product should be

$$K'_{S_{PbOx}} \geqslant [Pb^{2+}] [Ox^{2-}]$$

$$\geqslant 10^{-3} \times 9 \times 10^{-3}$$

$$\geqslant 9 \times 10^{-6}$$

This can only be achieved by complexing the lead with EDTA. This gives

$$K'_{S_{PbOx}} \geqslant K_{S_{PbOx}} \alpha_{Pb(EDTA)} \alpha_{Ox(H)}$$

so at pH 3.2

$$\alpha_{Pb(EDTA)} \geqslant 9 \times 10^{-6}/3 \times 10^{-11} \times 4.5$$

$$\geqslant 6.7 \times 10^4$$

That is, $$\alpha_{Pb(EDTA)} \geqslant 1 + K_{PbEDTA} [EDTA]$$

$$\geqslant 1 + K_{PbEDTA} \frac{[EDTA']}{\alpha_{EDTA(H)}}$$

where $[EDTA']$ is the total uncomplexed EDTA concentration (cf. footnote, p. 57) and

$$\alpha_{EDTA(H)} = 2 \times 10^{10} \text{ at pH } 3.2.$$

Therefore $[EDTA'] \geqslant 6.7 \times 10^4 \times 2 \times 10^{10}/8 \times 10^{17} \geqslant 1.7 \times 10^{-3}M$. The total EDTA concentration required is therefore $1.7 \times 10^{-3} + 1 \times 10^{-3} = 2.7 \times 10^{-3}M$. There is still the question of the possible complexation of the calcium by EDTA. Under the conditions used,

$$\alpha_{Ca(EDTA)} = 1 + K'_{CaEDTA}\,[EDTA']$$

$$= 1 + K_{CaEDTA}\,[EDTA']/\alpha_{EDTA(H)}$$

$$= 1 + 4 \times 10^{10} \times 1.7 \times 10^{-3}/2 \times 10^{10}$$

$$\approx 1$$

so there is no effect from this source. Here K'_{CaEDTA} is the conditional stability constant of the Ca–EDTA complex for a given pH (conditional because of protonation of the EDTA), cf. K'_S (p. 57).

The selectivity of the reactions discussed above has been considered solely on the basis of complex stability, and the effect of reaction conditions thereon. When the reaction involves the formation of a precipitate, the solubility of the complex in water must be considered. Also, in the case of the extraction of a complex into an organic solvent the relative solubilities of the complex in the aqueous and organic phases are important.

For example, the precipitation of a tervalent metal ion 8-hydroxyquinolinate can be represented by

$$M^{3+} + 3C_9H_6NOH \rightleftharpoons M(C_9H_6NO)_{3aq} + 3H^+$$

$$M(C_9H_6NO)_{3aq} \rightleftharpoons M(C_9H_6NO)_3(s)$$

The molar concentration of dissolved, undissociated metal 8-hydroxyquinolinate is its intrinsic solubility, S_i [56]. This can be calculated for any compound ML_m in equilibrium with its saturated solution in pure water, as follows:

$$M^{n+} + mL \rightleftharpoons ML_m$$

$$[M^{n+}]\,[L^-]^m = K_S \tag{2.98}$$

$$\frac{[ML_{m\ aq}]}{[M^{n+}]\,[L^-]^m} = \beta_m = \frac{S_i}{K_S} \tag{2.99}$$

The values of pK_S and pS_i for some metal ion 8-hydroxyquinolinates are given in Table 2.4. It can be seen that the intrinsic solubility varies much less than the solubility product for the different complexes. Such a small variation is expected

Table 2.4

Solubilities of some metal ion 8-hydroxyquinolinates [57]

Ion	pK_S	pS_i	Ionic radius (Å)	Covalent radius of metal (Å)
Mg^{2+}	15.6	6.7	0.65	1.36
Ca^{2+}	11.2	5.1	0.99	1.74
Sr^{2+}	9.3	4.8	1.13	1.91
Ba^{2+}	8.5	4.9	1.35	1.98

if the complexes are regarded as metal ions shielded by the same bulky, organic ligands. For a series of complexes of various metals with a given ligand in which the configuration is the same, the only structural difference will be the size, nature and charge of the metal ion. The size (for a given charge) should make little difference to the size and shape of the complex, and hence to the crystal lattice energy, the hydration energy, and thus the intrinsic solubility. As S_i is thus reasonably constant, Eq. (2.99) shows that the solubility product is inversely proportional to the stability constant, β_m, of the complex.

A similar argument can be used to show that the distribution coefficient for the extraction of complexes of similar structure increases with the stability constant of the complex.

Stability of Complexes with Organic Ligands

The following are the main factors which determine the stability of complexes with organic ligands

(1) *Ligand basicity.* The ability of the ligand to donate its lone pair of electrons to form a protonated species is a measure of its basicity. Within a series of similar ligands, this basicity is paralleled by the ability to donate electrons to form a complex with a particular metal ion.

(2) *Nature of the ligand.* Organic ligands which bond to a metal ion have electronegative centres through which bonding occurs. These are usually nitrogen (present as amino, diazo, nitroso or oximino groups), oxygen (present as hydroxy, carbonyl or carboxyl groups), and sulphur (present as thiol, thioketo or thiocarboxyl groups). The electronic structure of a metal ion determines its affinity for each type of donor. Ions having an 'inert gas' structure (sometimes called hard acids) form their strongest complexes with ligands which bond through oxygen atoms. Such ions include the alkali and alkaline earth metals, beryllium, magnesium, aluminium and thorium. Metal ions which have their d-orbitals partly filled, such as copper(II), iron(III), nickel(II) and cobalt(II), complex quite strongly through sulphur, oxygen and nitrogen. As the number of d-electrons increases, the affinity for sulphur increases compared with that for

oxygen. This simple concept can be used to predict, for example, which elements can be precipitated as their sulphides.

The concept of metal ions as 'hard' or 'soft' acids, and of ligand donor atoms as 'hard' or 'soft' bases is useful in this respect [58]. Its principles are embodied in Pauling's electroneutrality theory [59], namely that only a limited charge can build up on any one atom in a molecule. Thus, for a stable metal ion–ligand bond to be formed, the electronegativities of the acceptor and donor should be similar. Alternatively, on the basis of Fajans's rules of polarization [60], the polarizing effect of the cation should approximately match the polarizability of the donor atom. Otherwise excessive or minimal charge transfer occurs, with only weak bond formation.

Thus soft acids (weakly electronegative, weakly polarizing) such as mercury(II) and silver(I) form strong bonds with soft bases (weakly electropositive, readily polarized) such as sulphide ions or organic thiolates and other organosulphur compounds, whereas hard acids (more strongly electronegative, strongly polarizing) such as aluminium(III) and beryllium(II) bond strongly with hard bases (more strongly electropositive, difficultly polarized) such as fluoride ions, hydroxide ions and organic alkoxides.

(3) *Formation and size of chelate ring.* Chelates (p. 31) are more stable than the complexes formed with an equivalent number of otherwise similar unidentate ligands, that is ligands with only one donor atom. This effect arises partly because one multidentate ligand replaces at least two co-ordinated ligands (usually water in aqueous solution) from the metal ion, so that the disorder and therefore the entropy of the system is increased.

For example, the reactions of ammonia (unidentate) and ethylenediamine (en, bidentate) with a doubly-charged metal cation can be compared:

$$M(H_2O)_6^{2+} + 2NH_3 \rightleftharpoons M(NH_3)_2(H_2O)_4^{2+} + 2H_2O$$

$$M(H_2O)_6^{2+} + en \rightleftharpoons M(en)(H_2O)_4^{2+} + 2H_2O$$

In the first of these reactions there is no net change in the number of molecules in solution, whereas in the other one additional molecule is formed, thus increasing the entropy.

The equation for the change in the Gibbs free energy, ΔG, is $\Delta G = \Delta H - T\Delta S$ where ΔH is the enthalpy change, T is the absolute temperature and ΔS is the entropy change. Provided that the values of ΔH are similar for the reactions with the unidentate and multidentate ligands (true for ligands with the same donor atom and similar molecular structures), the chelating reaction, which causes ΔS to increase, gives a greater decrease in free energy, that is, gives a more stable complex.

This conclusion is supported by the values of the stability constants [61] of $M(NH_3)_2^{2+}$ and $M(en)^{2+}$ given in Table 2.5. Strictly, the stabilities of methylamine and ethylenediamine complexes should be compared, but as only a few methylamine complexes have been examined and their stabilities differ little from those of the corresponding ammonia complexes, the comparison between the stabilities of ammonia and ethylenediamine complexes serves to illustrate this point.

Table 2.5
Stability constants for some M^{2+} – ammonia
and ethylenediamine complexes

	Cd^{2+}	Zn^{2+}	Cu^{2+}	Ni^{2+}
$\log \beta_2$ (NH_3)	4.7	4.7	7.7	5.0
$\log \beta_1$ (en)	5.6	5.9	10.7	7.8

(Note that because the complexes are $M(NH_3)_2^{2+}$ and $M(en)^{2+}$ respectively, the stability constants should not be compared directly (because they have different units), but in terms of the concentration of free metal ion in equilibrium with a $1M$ concentration of complex).

Chelates are more stable when the bond angles of the ligand are less disturbed on chelation. This means that the most stable chelates are usually those with five or six members in the chelate ring, as in the complexes of 8-hydroxyquinoline (five), 1,10-phenanthroline (five), acetylacetone (six) and most other common chelating agents with metal ions (M^{n+})

| 8-hydroxyquinolinate | 1,10-phenanthroline | acetylacetonate |
| complex | complex | complex |

(4) *Nature of the metal ion.* The strength of a co-ordinate bond between a metal ion and a ligand bonding to oxygen or nitrogen increases with increasing charge density on the metal ion. Thus complexes of iron(III) with ligands such as tartrate or EDTA are more stable than the corresponding iron(II) complexes. Similarly, zirconium(IV) complexes are usually more stable than those of calcium(II) as a direct consequence of the greater electrostatic attraction of the

higher charge density (hard and soft acids). Cations of the '18-electron' type are more strongly polarizing than 'inert gas' type cations, and form more stable complexes.

Because both the ionic charge and size are important in determining the charge density, smaller ions with smaller charges, such as Be^{2+}, can form complexes of similar stability to those of larger more highly charged ions, such as Th^{4+}.

This simple concept does not apply to ligands with the heavier donor atoms, such as sulphur and phosphorus. In such instances, singly charged ions with low charge densities [silver(I), thallium(I)] often form stronger complexes than multicharged species of higher charge density [copper(II), iron(III)].

The number and spatial distribution of co-ordinate bonds to a metal ion are well defined. Cobalt(III), for instance, forms six co-ordinate bonds, disposed at right angles to each other to form an **octahedral** complex, as do Al(III), Fe(II), Fe(III), Ni(II), Zn(II) and Pb(II). Other metal ions form complexes with different structures and co-ordination numbers. Ag(I) and Hg(I) often form linear complexes with two ligands, whereas Cu(II) commonly forms a planar, 4-co-ordinate complex. Some of these are illustrated in Fig. 2.5.

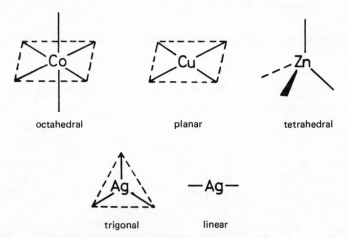

Fig. 2.5 – Spatial distribution of co-ordinate bonds.

Some of the heavier, multivalent metals, such as W(VI), Zr(IV), and U(VI), have co-ordination numbers of eight or more [62]. A number of metal ions form complexes having a variety of structures, depending on the nature of the ligand. Thus Ni(II) forms octahedral or planar structures with ligands bonding through oxygen or nitrogen, but tetrahedral complexes with some ligands bonding through sulphur; Ag(I) often forms tetrahedral or trigonal complexes with ligands bonding through heavy atoms (for example, sulphide ions, iodide ions, thiourea and phosphine).

Functional Groups

Ligands, especially chelating ligands, having particular donor atoms disposed in a certain molecular environment, often form analytically useful complexes with only a very restricted range of metal ions. For instance, reagents such as 1,10-phenanthroline and 2,2′-bipyridyl, which contain the grouping

form coloured complexes only with iron(II) and copper(I). This grouping, therefore, can be considered as a **functional group** for these species. Many other metal ions also form strong complexes with such reagents, but these complexes are colourless or only very weakly coloured. Thus, such reagents can be used to detect iron(II) specifically, although they can be used very effectively as masking agents for other metal ions.

Similarly, most compounds containing the α-dioxime grouping

such as dimethylglyoxime, precipitate only nickel, bismuth, palladium and platinum(II). Many other ions form *soluble* complexes, but the precipitation is diagnostic of the metals named, so this grouping is a functional group for the precipitation of nickel, palladium, platinum(II) and bismuth.

2.6 VISUAL EFFECTS IN FLAMES [63, 64]

Many inorganic salts, especially those of the alkali and alkaline earth metals, impart typical colours to the bunsen flame. This occurs because the flame is both hot enough to volatilize some of the salt and at the same time to dissociate it into atoms and/or radicals. The flame also provides enough energy to excite a small proportion of the metal atoms. When their excited electrons fall back to the ground-state, light of a characteristic wavelength is emitted. For example, sodium chloride boils at 1480°, so volatilization occurs in the bunsen flame. The $3s$ electron of a sodium atom is excited to a $3p$ orbital, and when it returns to the $3s$ level, light of wavelength *ca.* 590 nm is emitted, giving the typical yellow colour of sodium in a flame. The emission is a doublet of wavelengths 589.0 and 589.6 nm, because of electron spin effects.

All the alkali metal salts volatilize easily in a bunsen flame; these metals can readily be detected in many mixtures. However, the halides (other than the fluorides) of the alkaline earth metals and copper produce more intense flame colours than most other salts of these metals. This arises because many salts, such as the sulphates, decompose to give the metal oxide, which is difficult to volatilize in the flame. Addition of hydrochloric acid to the salts enhances the colour in the flame. Although the alkaline earth metals and copper produce atoms in the bunsen flame, the colours produced are at least partly due to excited molecular species. For example, the main calcium atomic emission is at 423 nm, but the red colour arises from emission due to CaOH, at 622 and 554 nm. Similarly, the green colour of copper flames is due to the emission from CuOH, CuO, and CuH, and, when chloride is present, CuCl. The green emission from boron arises from BO_2 species, and the red lithium emission arises partly from LiOH molecules.

Unlike the other elements mentioned, the alkaline earth metals do not emit radiation in the bunsen flame when certain other species are present. For example, phosphate and silicate ions form salts with the alkaline earths which are practically undecomposed in the flame, so emission does not occur to any significant extent. Similarly, aluminium and zirconium form very stable mixed oxides — spinels — with the alkaline earth metals, e.g.

$$2Al^{3+} + Ca^{2+} + 8NO_3^- \longrightarrow CaAl_2O_4(s) + 8NO_2(s) + 2O_2(g)$$

so again flame emission is suppressed. This effect can sometimes be nullified by adding a 'releasing' agent which forms an even more stable compound than does the metal of interest. Thus lanthanum will 'release' strontium in the presence of phosphate.

The use of a flame produced by burning hydrogen in air (hydrogen diffusion flame) gives rise to a different range of colours. Sulphur-containing ions, for example, because of the relatively low temperature, produce S_2 molecules, in an excited state, rather than sulphur atoms. When the excited electrons return to the ground state, the characteristic blue band emission (maximal intensity at *ca.* 384 nm) is produced. Similarly, phosphorus-containing species give rise to a green emission (maximal intensity *ca.* 520 nm) from HPO radicals. Many other non-metals will give similar emissions if they can be volatilized by or into a low-temperature hydrogen flame.

REFERENCES

[1] Clifford, A. F., *Inorganic Chemistry of Qualitative Analysis,* Prentice-Hall, London, 1961.
[2] Belcher, R., Nutten, A. J. and Stephen, W. I., *J. Chem. Soc.,* **1953**, 1334.
[3] Hahn, F. L. and Leinbach, G., *Ber.,* 1922, **55**, 3070.

[4] Guldberg, C. M. and Waage, P., *J. Prakt. Chem.,* 1879, **19**, 69.

[5] Sillén, L. G. and Martell, A. E., *Stability Constants of Metal-Ion Complexes,* Chem. Soc., London, 1964 (1st Supplement, Chem. Soc., 1971, 2nd Supplement, Pergamon, Oxford, 1979).

[6] Perrin, D. D., *Dissociation Constants of Organic Bases in Aqueous Solutions,* Butterworths, London, 1965; *Dissociation Constants of Inorganic Acids and Bases in Aqueous Solution,* Butterworths, London, 1969.

[7] Serjeant, E. P. and Dempsey, B., *Ionisation Constants of Organic Acids in Aqueous Solution,* Pergamon, Oxford, 1979; includes indexes to Kortüm, G., Vogel, W. and Andrussow, K., *Dissociation Constants of Organic Acids in Aqueous Solutions,* Butterworths, London, 1961.

[8] Lewis, G. N. and Randall, M., *J. Am. Chem. Soc.,* 1921, **43**, 1121.

[9] Debye, P. and Hückel, E., *Phys. Z.,* 1923, **24**, 185.

[10] Butler, J. N., *Ionic Equilibrium, A Mathematical Approach,* Addison-Wesley, Reading, Mass., 1964.

[11] Dyrssen, D., Jagner, D. and Wengelin, F., *Computer Calculation of Ionic Equilibria and Titration Procedures with Specific Reference to Analytical Chemistry,* Wiley, New York, 1968.

[12] Harned, H. S. and Owen, B. B., *Physical Chemistry of Electrolyte Solutions,* 3rd Ed., Reinhold, New York, 1958.

[13] Sørenson, S. P. L., *Compt. Rend. Lab. Carlsberg,* 1909, **8**, 1.

[14] Bishop, E., ed., *Indicators,* Pergamon, Oxford, 1972.

[15] Belcher, R. and Nutten, A. J., in association with Macdonald, A. M. G., *Quantitative Inorganic Analysis,* 3rd Ed., Butterworths, London, 1970.

[16] Smith, T. B., *Analytical Processes – A Physico-Chemical Interpretation,* 2nd Ed., Arnold, London, 1940.

[17] Van Slyke, D. D., *J. Biol. Chem.,* 1922, **52**, 525.

[18] Kolthoff, I. M. and Sandell, E. B., *Textbook of Quantitative Inorganic Analysis,* 3rd Ed., pp. 54–69, Macmillan, New York, 1952.

[19] Burns, D. T., *Ph.D. Thesis,* Leeds University, 1959.

[20] Ringbom, A., *Complexation in Analytical Chemistry,* Wiley, New York, 1963.

[21] Ref. [18], p. 59.

[22] Walton, A. G., *Formation and Properties of Precipitates,* Wiley, New York, 1967.

[23] Nielsen, A. E., *Kinetics of Precipitation,* p. 18, Pergamon, Oxford, 1964.

[24] Feigl, F. and Anger, V., *Spot Tests in Inorganic Analysis,* 6th Ed., p. 121, Elsevier, Amsterdam, 1972.

[25] Hamya, J. W. and Townshend, A., *Anal. Chim. Acta,* 1969, **46**, 312.

[26] Pröcke, O. and Uzel, R., *Mikrochim Acta,* 1938, **3**, 105.

[27] Gordon, L., Salutsky, M. L. and Willard, H. H., *Precipitation from Homogeneous Solution,* Wiley, New York, 1959.

[28] Salesin, E. D. and Gordon, L., *Talanta,* 1960, **5**, 81.

[29] Cartwright, P. F. S., Newman, E. J. and Wilson, D. W., *Analyst*, 1967, **92**, 663.

[30] Belcher, R., Farr, J. P. G. and Townshend, A., *Talanta*, 1969, **16**, 1089.

[31] Overbeek, J. Th.G., *Colloid Sci.*, 1952, **1**, 308.

[32] 17th IUPAC Conference, Stockholm 1953. See Licht, T. S. and de Béthune Pr. J., *J. Chem. Educ.*, 1957, **34**, 433 for a brief history of electrochemical sign conventions.

[33] For example, Latimer, W. M., *The Oxidation States of the Elements and their Potentials in Aqueous Solutions*, 2nd Ed., Prentice-Hall, Englewood Cliffs, N.J., 1952 (note that this book uses the opposite sign convention to IUPAC for electrode potentials).

[34] Morrison, G. H. and Freiser, H., *Solvent Extraction in Analytical Chemistry*, Wiley, New York, 1957.

[35] De, A. K., Khopkar, S. M. and Chalmers, R. A., *Solvent Extraction of Metals*, Van Nostrand-Reinhold, London, 1970.

[36] Marcus, Y. and Kertes, A. S., *Ion Exchange and Solvent Extraction of Metal Complexes*, Wiley-Interscience, New York, 1969.

[37] Starý, J., *Solvent Extraction of Metal Chelates*, Pergamon, Oxford, 1964.

[38] Hildebrand, J. H. and Scott, R. L., *Solubility of Non-Electrolytes*, 3rd Ed., Reinhold, New York, 1950.

[39] Bowd, A. J., Fogg, A. G. and Burns, D. T., *Talanta*, 1969, **16**, 719.

[40] Fogg, A. G., Burgess, C. and Burns, D. T., *Talanta*, 1971, **18**, 1175.

[41] Houben, J. and Fischer, W., *J. Prakt. Chem.*, 1929, **123**, 89. However, see also Friedman, H. L., *J. Am. Chem. Soc.*, 1952, **74**, 5.

[42] Green, H., *Talanta*, 1973, **20**, 139.

[43] West, P. W. and Mukherji, A. K., *Anal. Chem.*, 1959, **31**, 947.

[44] Chalmers, R. A. and Dick, D. M., *Anal. Chim. Acta*, 1964, **31**, 520; 1965, **32**, 117; Chalmers, R. A. and Svehla, G., in *Solvent Extraction Chemistry*, Dyrssen, D., Liljenzin, J.-O. and Rydberg, J., eds., p. 600, North-Holland, Amsterdam, 1967.

[45] Perrin, D. D., *Masking and Demasking of Chemical Reactions*, Wiley-Interscience, New York, 1970.

[46] Ref. [24], p. 421.

[47] Burnel, D., *Compt. Rend.*, 1965, **261**, 1982.

[48] Nisli, G. and Townshend, A., *Talanta*, 1968, **15**, 411.

[49] Ref. [27], p. 103.

[50] Perrin, D. D., *Organic Complexing Reagents*, Wiley-Interscience, New York, 1964.

[51] Holzbecher, Z., Diviš, L., Král, M., Šůcha, L. and Vláčil, F., *Handbook of Organic Reagents in Inorganic Analysis*, Horwood, Chichester, 1976.

[52] Burger, K., *Organic Reagents in Metal Analysis*, Pergamon, Oxford, 1973.

[53] Welcher, F. J., *Organic Analytical Reagents*, Vols. I-IV, Van Nostrand, New York, 1947.

[54] Burns, D. T., Townshend, A. and Carter, A. H., *Reactions of the Elements and their Compounds,* Horwood, Chichester, in preparation.

[55] Hutton, R. C., Salam, S. A. and Stephen, W. I., *J. Chem. Soc., A,* **1966,** 1573; **1967,** 1426.

[56] Irving, H. M. N. H. and Williams, R. J. P., *Analyst,* 1952, **77,** 813.

[57] Irving, H. M. N. H. and Rossotti, H. S., *Analyst,* 1955, **80,** 245.

[58] Pearson, R. G., *J. Am. Chem. Soc.,* 1963, **85,** 3533; *Science,* 1966, **151,** 172.

[59] Pauling, L., *The Nature of the Chemical Bond,* 3rd Ed., Cornell, Ithaca, 1960.

[60] Fajans, K., *Naturwiss.,* 1923, **II,** 165.

[61] Jones, M. M., *Elementary Coordination Chemistry,* p. 336, Prentice-Hall, Englewood Cliffs, N. J., 1964.

[62] Muetterties, E. L. and Wright, C. M., *Quart. Rev.,* 1967, **21,** 109.

[63] Gaydon, A. G., *The Spectroscopy of Flames,* 2nd Ed., Chapman & Hall, London, 1974.

[64] Alkemade, C. Th.J. and Hermann, R., *Fundamentals of Analytical Flame Spectroscopy,* Hilger, Bristol, 1979.

Chapter 3

Systematic Inorganic Qualitative Analysis-A Theoretical Interpretation

The consecutive separation of a mixture of substances into smaller and smaller groups, until it is possible to identify each component of a group, is referred to as systematic qualitative analysis. Usually it is mixtures of inorganic ions which are analysed in this way, and numerous schemes for this purpose have been proposed over the years [1]. Some of the early developments are discussed in Chapter 1. It has been pointed out that the process is analogous to that used in punched-card sorting [2].

Systematic qualitative analysis evolved because it was necessary to be able to detect ions in complicated mixtures, in the days when analytical instrumentation was extremely rudimentary, and the number of characteristic reactions was limited. Nowadays, powerful analytical equipment, such as the direct reading emission spectrometer, the X-ray fluorescence spectrometer and the mass spectrometer, enables complete qualitative analyses of complex samples, often in minute amounts, to be made in a matter of minutes. For occasional laboratory testing, however, or for rapid field testing, qualitative analysis retains a facility and economy still unsurpassed by modern instrumentation. Yet even in such circumstances, completely *systematic* analysis is usually unnecessary, because of the availability of sensitive, highly selective test reactions for most inorganic species.

Nevertheless, by utilizing the most effective reagents and reactions available, systematic qualitative analysis provides students with a comprehensive experience of experimental inorganic solution chemistry that cannot be achieved so effectively or economically within a relatively short time in any other way [3]. By consideration of the concepts and quantitative relationships involved in the wide variety of reactions encountered, the student is able to put into practice the theory and principles which he has been taught. Such experience and general chemical knowledge is often sadly lacking in new graduate chemists.

The main value of systematic qualitative analysis, therefore, is in *teaching*.

It fulfils several useful functions in this respect. The diligent student, in carrying out the wide range of reactions used, will experience the qualitative and quantitative application of those reactions and fundamental principles described in Chapter 2, which are of such great importance in current analytical chemistry. The use of small-scale equipment teaches the student to cultivate a precise and careful experimental technique. Finally, students will learn that to obtain acceptable separations and identifications, it is necessary to work in a clean and tidy environment, develop a critical observational facility and be able to evaluate observations logically.

In this book, a scheme of systematic qualitative analysis is presented that has been evolved by modification and innovation from previous schemes. It has been subjected to continuous detailed scrutiny and experiment over the last twenty years, and numerous improvements and extensions have been made in this time. Unlike many schemes, it includes systematic separations of cations *and* of anions. Final identification is made by tests that are considered to be the most reliable. In addition to the systematic schemes, there is a wide range of additional tests (for specific species, or with certain reagents) and a special scheme for dealing with a range of difficultly soluble compounds.

Detailed experimental instructions concerning this scheme are given in Chapters 4 and 5. In this chapter, the quantitative application of the physico-chemical principles outlined in Chapter 2 to the separations in the systematic schemes is given together with explanations of the chemistry of many of the tests and separations. Where these are omitted, the reader is referred to the companion book [4] which presents detailed accounts of characteristic reactions of very many inorganic ions and compounds, including all those encountered in Chapter 5.

3.1 THE MAQA SCHEME

Like other schemes of systematic qualitative inorganic analysis, the MAQA scheme first separates a complex mixture into smaller groups of cations or anions.

For the separation of cations, the traditional separations by precipitation as chlorides (Ag^+, Hg^{2+}, Pb^{2+}), sulphides from acidic solution [Hg^{2+}, Cu^{2+}, Bi^{3+}, Cd^{2+}, As(III), Sb(III), Sn(IV)], hydrated oxides (Fe^{3+}, Cr^{3+}, Al^{3+}) and sulphides from alkaline solution (Zn^{2+}, Mn^{2+}, Ni^{2+}, Co^{2+}) are made, in that order. In contrast to most other schemes, however, the alkaline earth metals are precipitated as their sulphates after the chloride precipitation. Usually, these ions are precipitated as their carbonates after the sulphide precipitation from alkaline solution. However, if that procedure is used, appreciable loss of alkaline earth metals as their sulphates may already have occurred, because of the formation of sulphate ions by oxidation of sulphide ions. Such loss often leads to considerable difficulty in identifying these metals by carbonate precipitation. Some of the cations remaining in solution after all these precipitations (Na^+, Li^+, Mg^{2+}) are treated as another group. Because of the possible accumulation of impurities by this stage, potassium is not tested for in this group, but during preparation of the sample for systematic anion analysis (p. 171), where little contamination is likely.

The scheme for anion separation is described on p. 192. It provides a rapid and reliable means of separating and identifying most, but by no means all, common inorganic, and some organic, anions. In order to make a complete anion analysis, therefore, it is necessary to carry out a number of preliminary general and specific tests in addition to the systematic scheme. The range of anions included is larger than in most other schemes, systematic or otherwise.

A scheme for the treatment and analysis of difficultly soluble substances is described on p. 213. This provides for the treatment of the sample with reagents of increasing vigour (if the normal acidic solvents have already proved to be unsuccessful), so that the resulting solutions can be systematically analysed. The scheme does not cover all 'insoluble' materials, but accommodates those most likely to be met with in undergraduate teaching classes, and illustrates the general principles involved.

The MAQA scheme has a further advantage over most other schemes. It has been extended to cover a large number of elements not normally accommodated by systematic schemes. These elements are W, Tl, Se, Mo, Te, V, U, Zr, Ti, Be, Th and Ce. Although the schemes have not been tested for every possible combination of these elements, most mixtures are readily separated and identified.

3.2 SYSTEMATIC CATION SEPARATION SCHEME

In the MAQA scheme, mixtures are first separated into groups of cations by selective precipitation reactions. These groups are identified by name to avoid confusion with the group numbers of earlier schemes. The alkaline earth metals, for example, are separated as the second group, here called the Calcium Group, whereas in the earlier schemes they were often separated later and termed Group V. The group names, their constituent cations and the order of separation are given below:

1. *Silver Group:* silver(I), mercury(I), lead(II), thallium(I), precipitated as chlorides, and tungsten(VI) as tungstic acid.

2. *Calcium Group:* calcium(II), strontium(II), barium(II) and lead(II) precipitated as sulphates.

3. *Copper – Tin Group:* copper(II), mercury(II), bismuth(III), cadmium(II), selenium(IV), selenium(VI), tin(II), tin(IV), antimony(III), arsenic(III), arsenic(V), tellurium(IV), tellurium(VI) and molybdenum(VI), precipitated by hydrogen sulphide from acidic solution.

4. *Iron Group:* iron(III), aluminium(III), chromium(III), zirconium(IV), titanium(IV), cerium(III), thorium(IV), uranium(VI), vanadium(IV), vanadium(V) and beryllium(II), precipitated by making the solution ammoniacal after the removal of hydrogen sulphide.

5. *Zinc Group:* zinc(II), manganese(II), nickel(II) and cobalt(II), precipitated as sulphides from alkaline solution.

6. *Magnesium Group:* magnesium(II), sodium(I) and lithium(I), which remain in solution.

In the following sections the basis of the separation of each group as a whole, and of the further separations which lead to the unequivocal identification of the constituent cations, is discussed. Where sufficient equilibrium data are available, and where equilibrium calculations are applicable, a quantitative interpretation of the separations is given.

These Groups of the more common cations are generally similar to those groups separated in other systematic schemes, although the order of separation and the precipitants used may be different. Although the present scheme for analysis of each Group is considered to be the best available, many others have been suggested. Where such alternative schemes of separation have been reviewed, the appropriate reference is given at the beginning of each Group. The elements in each Group are given at the beginning of each discussion. Those in parenthesis are less common elements, which are included in the extended version of the scheme.

3.2.1 Silver Group [5]

Silver(I), mercury(I), lead(II), [thallium(I), tungsten(VI)]

The addition of dilute hydrochloric acid to a solution containing silver(I), mercury(I), lead(II), and/or thallium(I) ions precipitates these ions as their chlorides. The cations (M^{n+}) will be precipitated until the ionic product for the ions remaining in solution is equal to the solubility product:

$$[M^{n+}] [Cl^-]^n = K_S$$

Precipitation then ceases and any cations remaining in solution appear in Groups separated subsequently. With a slight excess of chloride ion precipitation of mercury(I) and silver(I) chlorides is virtually complete, but lead(II) and thallium(I) chlorides are slightly soluble in cold water and much more soluble in hot water (cf. p. 100). It is, therefore, essential to cool the solution thoroughly after the addition of hydrochloric acid, before centrifuging. An appreciable amount of lead may remain in solution even after cooling, however, as is shown below, and it must be tested for again in the Calcium Group.

The use of the concentrations recommended in the separation scheme (p. 174) gives a solution $10^{-2}M$ with respect to each cation, assuming there are three different cationic species present. Because of the relatively imprecise method of measuring samples for qualitative analysis [6] this concentration may be larger or smaller by at least a factor of 3; this should be noted when considering the implications of the calculations made below.

After addition of the dilute hydrochloric acid, the chloride ion concentration is approximately $1M$. The ionic concentration product for a metal chloride, MCl, is therefore $10^{-2} \times 1 = 10^{-2}$ mole^{2}l^{-2}, and that for MCl$_2$ is $10^{-2} \times 1^2 = 10^{-2}$ mole^3.l^{-3} (units will henceforth be omitted; they are readily calculated). Only metal chlorides with solubility products appreciably smaller than these values will precipitate under these conditions (Table 3.1).

Table 3.1

Solubility products (25°C) of some chlorides

Chloride	K_S
AgCl	2×10^{-10}
TlCl	2×10^{-4}
CuCl	1×10^{-6}
(Hg$_2$)Cl$_2$	2×10^{-18}
PbCl$_2$	2×10^{-5}
HgCl$_2$	6×10^{-2}

Any copper(I) present in the original mixture should disproportionate to copper metal and copper(II) when the solution for cation analysis is prepared even in water or dilute hydrochloric acid, unless it is present as a stable complex or an insoluble compound, but may be aerially oxidized to copper(II). Copper(II) chloride is readily soluble in water, so copper will not precipitate in this Group. If an oxidizing solvent is used, mercury(I) will be oxidized to mercury(II), which does not precipitate as its chloride. Of the other chlorides in Table 3.1, lead and thallium have the largest solubility products, and are therefore the most soluble in dilute hydrochloric acid or in water.

The amount of lead remaining in the solution can readily be calculated. The chloride ion concentration is $1M$, thus

$$K_S = [\text{Pb}^{2+}] [\text{Cl}^-]^2 = [\text{Pb}^{2+}] \times 1^2 = 2 \times 10^{-5}$$

$$[\text{Pb}^{2+}] = 2 \times 10^{-5}M$$

That is, the concentration of lead ions remaining in solution is $2 \times 10^{-5}M$, (4 μg/ml).

Another factor that influences the solubility of lead ions, but is often ignored though it should be taken into account, is that lead ions form a series of complexes with chloride ions, of general formula $[\text{PbCl}_n]^{(2-n)+}$, $n = 1, 2, 3$, or 4. The complex ion PbCl$_4^{2-}$ is extremely unstable, so PbCl$_3^-$ is the highest complex formed. This complex anion has a stability constant (β_3) of about 20.

As the concentration of chloride ions is high relative to the lead ion concentration, it can be shown that lower complexes do not contribute significantly to the total lead concentration in solution. Thus:

$$\beta_3 = \frac{[PbCl_3^-]}{[Pb^{2+}][Cl^-]^3} = 20$$

The lead ion concentration is therefore reduced by complex formation, to a value which may be calculated as follows. If the total concentration of lead species is $[Pb]_t$, then

$$[PbCl_3^-] = [Pb]_t - [Pb^{2+}]$$

$$= 10^{-2} - [Pb^{2+}]$$

Substituting this into the value for β_3 gives

$$\frac{10^{-2} - [Pb^{2+}]}{[Pb^{2+}] \times 1^3} = 20$$

so $$[Pb^{2+}] = 5 \times 10^{-4} M$$

Such a lead ion concentration gives an ionic concentration product of lead chloride of $5 \times 10^{-4} \times 1^2 = 5 \times 10^{-4}$, which is appreciably greater than the solubility product of lead chloride, so precipitation will occur. However, some lead will remain in solution. The amount remaining can be calculated by using β_3. After precipitation, the lead ion concentration must be $2 \times 10^{-5} M$, from the solubility product expression. Substituting into the value for β_3 gives

$$[PbCl_3^-] = 2 \times 10^{-5} \times 1^3 \times 20$$

$$= 4 \times 10^{-4} M$$

Thus the total concentration of unprecipitated lead in solution is

$$[Pb^{2+}] + [PbCl_3^-] = 2 \times 10^{-5} + 4 \times 10^{-4} \sim 4 \times 10^{-4} M ,$$

and not the $2 \times 10^{-5} M$ calculated without taking complex formation into account. Greater chloride ion concentrations increase the solubility significantly. In practice, precipitation equilibrium is unlikely to be reached, so the lead ion concentration will be appreciably greater than $4 \times 10^{-4} M$. The lead unprecipitated at this stage is precipitated in the following Calcium Group, where the detection of lead in the Silver Group can be confirmed.

The amount of thallium remaining in solution can similarly be calculated from the solubility product expression:

$$K_S = [Tl^+][Cl^-] = [Tl^+] \times 1^2 = 2 \times 10^{-4}$$

$$[Tl^+] = 2 \times 10^{-4} M$$

Thallium(I), like mercury(I), has little tendency to form higher chloro-complexes, so these make a negligible contribution to the solubility. The amount of thallium unprecipitated at equilibrium, therefore, is about half that of lead. Any unprecipitated thallium is eventually precipitated in the Zinc Group. Silver(I) forms a soluble complex, $AgCl_2^-$ ($\beta_2 \cong 10^5$). However, because of the relative insolubility of silver chloride an insignificant amount of silver remains unprecipitated. This can be shown by using calculations similar to those for lead chloride.

If silicate and tungstate ions are present the addition of hydrochloric acid precipitates the hydrated oxides of silicon(IV) and tungsten(VI). Niobium, tantalum and titanium behave similarly, but precipitation is less complete with these metals. The tungsten precipitate is accommodated in the subsequent separation of the Silver Group ions. The other ions are not, but they do not interfere unduly with subsequent operations and tests.

Some other compounds may also precipitate in the Silver Group. Barium chloride and, less likely, sodium chloride may precipitate, but only if the cation and hydrochloric acid concentrations are both extremely high. They dissolve on dilution with water. Compounds of bismuth and antimony may hydrolyse to give precipitates of the oxide chlorides, for example, $Bi^{3+} + Cl^- + H_2O \rightleftharpoons BiOCl(s) + 2H^+$. This reaction can be prevented by increasing the concentration of hydrochloric acid. Borate ions and anions of some organic acids such as benzoic and salicylic are protonated and the acids may precipitate:

$$BO_3^{3-} + 3H^+ \longrightarrow H_3BO_3(s)$$

$$C_6H_5COO^- + H^+ \longrightarrow C_6H_5COOH(s)$$

Such anions should be removed (before the hydrochloric acid is added) by addition of dilute nitric acid to precipitate the acids. The solution must then be evaporated to remove the nitric acid before proceeding to the Calcium Group precipitation. Finally, if thio-salts of arsenic, antimony or tin are present, the appropriate metal sulphide precipitates on acidification. This is again avoided by acidifying first with nitric acid, although the formation of sulphate ions may precipitate some Calcium Group ions. Thus any precipitate formed should be tested for Calcium Group ions.

If Cu^{2+} and I^- are both present, CuI will be precipitated and iodine released (recognizable by violet vapour on warming, or by testing with starch).

Separation of the Chloride Precipitate (pp. 179, 205)

Solubility *vs* temperature data show that lead(II) and thallium(I) chlorides readily dissolve in boiling water, and may thus be separated from silver(I) and mercury(I) chlorides and from hydrated tungsten(VI) oxide (tungstic acid).

The *solution* so obtained is treated with sulphuric acid, and evaporated until fumes of sulphuric acid appear. This volatilizes chloride ions as hydrogen chloride, and forms lead(II) and thallium(I) sulphates. On dilution of the cooled suspension, however, the thallium sulphate, which is appreciably soluble, dissolves, leaving a precipitate of lead sulphate. Lead and thallium are tested for in the precipitate and solution, respectively.

The *residue* which remains after the hot water extraction is treated with an excess of ammonia. A calculation like that given on p. 76 shows that the solubility of silver chloride in $4M$ ammonia solution is $0.1M$. As the metal ion concentrations normally used in this scheme are $10^{-2}M$, all the silver chloride precipitated from 1 ml of sample solution dissolves readily in 1 ml of $4M$ ammonia solution. Tungstic acid also dissolves in the alkaline solution as tungstate ions. Mercury(I) chloride, however, is converted by disproportionation into a black residue containing elemental mercury and mercury(II) chloride.

Tungstate and silver ions are separated by adding iodide ions to precipitate silver iodide. Assuming the concentration of silver ammine in the 2 ml of solution is $5 \times 10^{-3}M$ at this stage (p. 96), and the ammonia concentration after pH adjustment is $0.1M$, then the concentration of silver ions is given by Eq. (2.56):

$$[Ag^+] = \frac{[Ag(NH_3)_2^+]}{\beta_2[NH_3]^2} = \frac{5 \times 10^{-3}}{2 \times 10^7 \times (0.1)^2} = 2.5 \times 10^{-8}M$$

The iodide ion concentration, immediately after addition, is $10^{-2}M$. Thus the ionic concentration product

$$[Ag^+][I^-] = 2.5 \times 10^{-8} \times 10^{-2}$$

$$= 2.5 \times 10^{-10}$$

which considerably exceeds the solubility product of silver iodide, 10^{-16}, so silver iodide is precipitated. Iodide ions do not react with tungstate ions under these conditions. Tungsten and mercury are tested for in the ammoniacal solution and the black precipitate respectively. The yellow silver iodide precipitate is considered to be suffcient identification of silver.

If tungsten is not included in the scheme, alternative confirmatory tests may be applied to the silver ammine solution.

3.2.2 Calcium Group [7]

Calcium, strontium, barium, lead(II)

The solution remaining after the removal of the Silver Group is treated with dilute sulphuric acid; barium, strontium, calcium and lead(II) ions (if present) precipitate as their sulphates. Precipitation continues until the ionic concentration product equals the solubility product:

$$[M^{2+}] \, [SO_4^{2-}] = K_S$$

As calcium and lead sulphates are appreciably soluble in water, ethanol is added to reduce their solubilities and obtain a more complete precipitation. However, the amount of ethanol added must be carefully controlled. If too much is added other cations may be precipitated; if insufficient is added, precipitation may be incomplete, and calcium sulphate, which is slow to nucleate, may not precipitate.

If lead has been found in the Silver Group, it will also be present in the Calcium Group because of the incomplete precipitation of its chloride (p. 97). However, if lead has *not* been found in the Silver Group, it may still be found in the Calcium Group, either because of its low concentration or because inadequate cooling of the solution prevented its precipitation in the Silver Group.

To understand the basis of the separation of the Calcium Group cations the sulphate ion concentration in the test solution must first be calculated. Sulphuric acid is completely dissociated in its first dissociation stage, but the subsequent dissociation of the hydrogen sulphate ion is incomplete (p. 40).

$$K_2 = \frac{[H^+] \, [SO_4^{2-}]}{[HSO_4^-]} = 1 \times 10^{-2} \tag{3.1}$$

The sulphuric acid concentration after addition to the test solution is $0.2M$. Assuming that the sulphate ion concentration is small compared to that of the hydrogen sulphate ion, $[HSO_4^-] \cong 0.2M$, the total hydrogen ion concentration is effectively that given by the hydrogen ions released by the first stage dissociation of sulphuric acid together with the hydrogen ion concentration present before the addition of the sulphuric acid (that is from the hydrochloric acid added to precipitate the Silver Group). The hydrogen ion concentration resulting from the dissociation of the hydrogen sulphate ion is negligible. Thus, taking into account the dilution of the solution after precipitation of the Silver Group:

$[H^+]$ from sulphuric acid $\cong 0.2M$

$[H^+]$ from hydrochloric acid $\cong 0.4M$

$[H^+]$ total $\cong 0.6M$

Substituting these values into Eq. (3.1) gives

$$[SO_4^{2-}] = \frac{1 \times 10^{-2} \times 0.2}{0.6} \sim 3 \times 10^{-3}M$$

which confirms the assumption that the sulphate ion concentration is much smaller than that of the hydrogen sulphate ion. Assuming the solution to be $3 \times 10^{-3}M$ with respect to the cation to be precipitated, because the solution has been diluted about threefold since it was initially prepared, the ionic concentration product is

$$[M^{2+}] [SO_4^{2-}] = 3 \times 10^{-3} \times 3 \times 10^{-3} \cong 10^{-5}$$

Only metal sulphates with solubility products appreciably less than this value will precipitate under these conditions. The solubility products for such sulphates in aqueous solutions are given in Table 3.2. The precipitation is carried out, however, in 50% aqueous ethanol, and the solubility products in this medium do not appear to have been measured reliably. As the solubilities of these salts are reduced by about two orders of magnitude in the change from water to 50% aqueous ethanol, the solubility products should be lower by a factor of $\sim 10^4$ in the latter medium. Approximate values calculated on this basis are included in Table 3.2.

Table 3.2
Solubility products (25°C) of some sulphates

Sulphate	K_S (H$_2$O)	Approx. K_S (1:1 H$_2$O – ethanol)
BaSO$_4$	1×10^{-10}	10^{-14}
SrSO$_4$	5×10^{-7}	10^{-10}
CaSO$_4$	4×10^{-5}	10^{-8}
PbSO$_4$	2×10^{-8}	10^{-12}

The values in Table 3.2 show that barium, strontium and lead sulphates should precipitate in the aqueous or aqueous ethanolic medium. The ion-product for calcium sulphate, under the test conditions, however, appears just not to exceed the solubility product for wholly aqueous media, but does exceed that suggested for the ethanol-containing solution. Thus the presence of ethanol is necessary to ensure precipitation of an appreciable amount of the calcium present, and also to minimize the slow nucleation and persistent supersaturation effects that delay calcium sulphate precipitation in aqueous media. Even when ethanol is present, the solution should be well stirred, and about 5 minutes allowed for the precipitate to form.

Separation of the Sulphate Precipitate
The precipitate is first treated with boiling water. The solubilities of the relevant sulphates in boiling water are given in Table 3.3.

Table 3.3
Solubilities of some sulphates in boiling water [8]

	$CaSO_4$	$SrSO_4$	$BaSO_4$	$PbSO_4$
Solubility of metal ion, mg/ml	0.51	0.06	0.0025	0.07

As 1.5 ml of water are used, 0.8 mg of calcium ions should be dissolved, giving a $1.3 \times 10^{-2}M$ solution. In theory this means that all the calcium sulphate should be extracted from the precipitate by the boiling water, as the original 1 ml of solution is assumed to have only $10^{-2}M$ concentration of calcium. In contrast, the extract of strontium sulphate is only $6.8 \times 10^{-6}M$, and that of lead sulphate $3.4 \times 10^{-4}M$. Thus calcium is appreciably separated from the other metal ions by dissolution in boiling water, and may be tested for in the aqueous extract. However, it should be remembered that calcium sulphate co-precipitated with other metal sulphates will be less readily extracted than calcium sulphate alone [7], so in practice some calcium will remain in the precipitate and will accompany strontium through the subsequent separation stages.

The subsequent treatment of a mixed sulphate precipitate, possibly containing strontium, barium and lead, has caused much difficulty for analytical chemists, mainly because of the difficulty of dissolving barium sulphate. In the present scheme two procedures are given: one based on metathesis with sodium carbonate and dissolution of the resulting carbonates in acetic acid, and the other based on the dissolution of the sulphates by EDTA. The metathesis, which has been in use for many years, gives reasonable results, but is time-consuming and for more rapid working the EDTA separation is preferable. Both schemes are described below.

(a) *Metathesis of sulphate precipitate*
The metal sulphates undissolved by boiling water are converted into their respective insoluble metal carbonates by metathesis with 2.5 ml of $1M$ sodium carbonate solution:

$$MSO_4(s) + CO_3^{2-} \longrightarrow MCO_3(s) + SO_4^{2-}$$

The conversion into carbonates is carried out because the carbonates are readily dissolved in acid (in this scheme dilute acetic acid is used).

The effectiveness of the metathesis depends on the relative solubilities of the metal sulphate and carbonate, and on the carbonate ion concentration. For example, in the metathesis of barium sulphate, the ratio of the solubility products at room temperature is:

$$\frac{[Ba^{2+}][CO_3^{2-}]}{[Ba^{2+}][SO_4^{2-}]} = \frac{5 \times 10^{-10}}{1 \times 10^{-10}} = 5$$

At equilibrium, the barium concentration is the same in both numerator and denominator, so after the mixture has cooled to room temperature,

$$\frac{[CO_3^{2-}]}{[SO_4^{2-}]} = 5$$

The initial carbonate ion concentration will be decreased by an amount equal to the concentration of sulphate released, so the maximal sulphate ion concentration must be a sixth of the initial carbonate concentration, that is 4×10^{-4} mole in the 2.5 ml of solution. As the sulphate can only have been released by conversion of barium sulphate into barium carbonate, this indicates that as much as 4×10^{-4} mole (ca. 56 mg) of barium may be converted into the carbonate in this way. Thus, in theory, all the barium sulphate (under the conditions of the scheme) should be converted into barium carbonate.

The corresponding solubility product ratios for the other metal carbonates and sulphates are much smaller, and conversion of the sulphates into the carbonates would also be essentially complete. These reactions, however, are slow, and in practice, even after boiling for 5 minutes conversion is incomplete, largely because there will be a protective coating of carbonate on the surface of the sulphate. After the carbonate precipitate has been dissolved in acetic acid, any residue will be unconverted sulphate precipitate, which is discarded. The amount of residue, however, gives a measure of the effectiveness of the metathesis.

From the solution lead, if present, is first precipitated as sulphide, and barium is separated from strontium in the resulting solution by precipitation as chromate. This separation can be shown to be selective by calculating the concentrations of barium and strontium ions needed for precipitation of barium and strontium chromates, respectively, after the addition of potassium chromate: the volume of the solution after addition of potassium chromate to test for barium is ca. 0.8 ml, and initially has $0.1M$ total chromate concentration. Chromic acid is a weak acid, so in acidic solution the protonation of chromate will be extensive

$$HCrO_4^- \rightleftharpoons CrO_4^{2-} + H^+$$

The concentration of chromate ions can be calculated from the dissociation constant

$$K_2 = \frac{[CrO_4^{2-}] \, [H^+]}{[HCrO_4^-]} = 3 \times 10^{-7}$$

At any hydrogen ion concentration greater than $3 \times 10^{-7}M$, $[HCrO_4^-]$ will be the predominant species. The concentration of acetic acid, c, after dissolution of the carbonate precipitate, is about $2M$. As acetic acid is also a weak acid, $K_1 = 2 \times 10^{-5}$, the approximate hydrogen ion concentration is given by Eq. (2.19) as

$$[H^+] = \sqrt{K_1 C} = \sqrt{2 \times 10^{-5} \times 2} \sim 6 \times 10^{-3}M \; .$$

Hence $[CrO_4^{2-}]/[HCrO_4^-] = 3 \times 10^{-7}/6 \times 10^{-3} = 5 \times 10^{-5}$, so virtually all the chromate will be in the form $[HCrO_4^-]$. Hence $[CrO_4^{2-}] = 5 \times 10^{-5} \times 0.1 = 5 \times 10^{-6}M$.

The solubility product of barium chromate is 2×10^{-10}. Therefore, for barium chromate to precipitate under these conditions, the barium ion concentration must appreciably exceed

$$[Ba^{2+}] = \frac{K_S(BaCrO_4)}{[CrO_4^{2-}]} = \frac{2 \times 10^{-10}}{5 \times 10^{-6}} = 4 \times 10^{-5}M$$

If it is assumed that half the barium sulphate is brought into solution as a result of the metathesis and acid treatment, the barium ion concentration in the solution immediately on addition of the chromate is about $6 \times 10^{-3}M$, which is greatly in excess of that required, $4 \times 10^{-5}M$, to bring about barium chromate precipitation. The solubility product of strontium chromate is 4×10^{-5}. In order for strontium chromate to precipitate from the solution, the strontium concentration must appreciably exceed

$$[Sr^{2+}] = \frac{K_S(SrCrO_4)}{[CrO_4^{2-}]} = \frac{4 \times 10^{-5}}{5 \times 10^{-6}} = 8M$$

Obviously, this concentration is not exceeded and strontium is not precipitated. Therefore the solution contains the strontium and the barium is precipitated.

(b) *EDTA dissolution of sulphate precipitate*

One of the very few simple methods of dissolving barium sulphate in aqueous solution is to treat the precipitate with an alkaline solution of EDTA (H_4Y).

Barium ions are complexed by the EDTA anions (Y^{4-}), so the precipitate dissolves to give sulphate ions and the barium-EDTA complex, BaY^{2-}. An alkaline solution is necessary in order to prevent protons from competing successfully for the EDTA anion. As EDTA is a weak acid in all of its dissociation stages, protonated species such as HY^{3-} and H_2Y^{2-} predominate unless the solution is appreciably alkaline. Because of this protonation side-reaction, barium sulphate does not dissolve in EDTA if the pH is too low.

The sulphate precipitate is treated with 1 ml of 0.5M ammoniacal EDTA solution. It can be shown that all the precipitate dissolves. The competing equilibria are, for barium sulphate:

$$Ba^{2+} + SO_4^{2-} \rightleftharpoons BaSO_4(s) \qquad K_S = 1 \times 10^{-10}$$

$$Ba^{2+} + Y^{4-} \rightleftharpoons BaY^{2-} \qquad K = 6 \times 10^7$$

At pH 10, however, log $\alpha_{Y(H)}$ is 0.6, so we must use the conditional constant (p. 82, charges omitted for simplicity where necessary), $[BaY]/[Ba][Y'] = K' = K/\alpha_{Y(H)} = 1.5 \times 10^7$. Then $[BaY][SO_4^{2-}]/[Y'] = K_S/K' = 1 \times 10^{-10} \times 1.5 \times 10^7 = 1.5 \times 10^{-3}$. Since BaY is formed from $BaSO_4$ with liberation of SO_4^{2-}, $[BaY] \sim [SO_4^{2-}]$, and as excess of EDTA has been added then

$$[Y'] = [Y]_{added} - [BaY] \sim [Y]_{added}$$

so $\qquad\qquad [BaY]^2 \sim 1.5 \times 10^{-3} \times 0.5 \sim 7.5 \times 10^{-4}$

Thus [BaY], which is the maximal concentration obtainable by dissolving the precipitate in this way, is $2.7 \times 10^{-2}M$. As the concentration of barium taken is only $1 \times 10^{-2}M$, all of the barium sulphate formed should dissolve.

Strontium and lead, which form stronger EDTA chelates ($K = 5 \times 10^8$ and 10^{18} respectively) and more soluble sulphates than barium, can obviously also be completely dissolved in the EDTA solution.

Chromate (0.5 ml of 0.5M solution) is added to the 1 ml of the EDTA solution. No precipitation occurs because the metal ion concentrations are insufficient to exceed the appropriate solubility products. The unchelated metal ion concentration in the solution is given by

$$[M^{2+}] = \frac{[MY]}{[Y]K} \alpha_{Y(H)} \qquad \text{(where M = Ba, Sr or Pb)}$$

Using the assumptions made above, and remembering that the solution has been diluted from 1 ml to 1.5 ml,

$$[M^{2+}] = \frac{\frac{2}{3} \times 1 \times 10^{-2}}{\frac{2}{3} \times 0.3 \times K} \times 4M$$

and
$$[CrO_4^{2-}] = \frac{0.5}{3} = 0.17M$$

The concentrations for barium, strontium and lead are given in Table 3.4, as are the respective ionic concentration products. They show that the solubility products of barium, strontium and lead are not exceeded under these conditions.

Barium chromate is selectively precipitated from the EDTA–chromate solution by adding magnesium ions [9] (0.5 ml of 1.2M solution) to displace most of the barium ions from the EDTA chelate. Because of the greater strength of their chelates, insufficient lead or strontium will be released to cause significant precipitation of their chromates. The displacement reaction is:

$$MY^{2-} + Mg^{2+} \rightleftharpoons MgY^{2-} + M^{2+}$$

so
$$\frac{[MgY][M^{2+}]}{[MY][Mg^{2+}]} = \frac{K(Mg)}{K(M)} = K_{ex}$$

As $K(Mg) = 5 \times 10^8$, $K_{ex} = 8$ for M = Ba, 1 for M = Sr and 5×10^{-10} for M = Pb, and as the total concentrations of magnesium and EDTA are much larger than that of M, most of the EDTA will be bound as its magnesium chelate:

$$[MgY] \cong [Y]_{total} \cong 0.25M \text{ in the 2 ml of solution.}$$

Also, as the total magnesium concentration is 20% greater than the total EDTA concentration, it can be assumed that

$$[Mg^{2+}] = [Mg]_{total} - [Y]_{total} = 0.05M$$

Therefore $[M^{2+}]/[MY] = K_{ex} \times \dfrac{0.05}{0.25} = 0.2K_{ex}$ and also $[M^{2+}] + [MY] = 5 \times 10^{-3}M$, so $[M^{2+}] = 5 \times 10^{-3}M/(1 + 5/K_{ex})$.

Thus $[M^{2+}]$, and the ionic concentration product $[M^{2+}][CrO_4^{2-}]$, can be calculated. The values are given in Table 3.4. They show that the addition of the magnesium ions displaces 60% of the barium from its EDTA chelate, thus causing the solubility product of barium chromate to be greatly exceeded, and most of the barium present to precipitate as the chromate. The solubility product of barium sulphate is also exceeded but, because barium sulphate and chromate have similar solubilities, and because there is a large excess of chromate ions over sulphate ions, the precipitate is mostly barium chromate.

Table 3.4

Concentrations and concentration products relevant to the EDTA/chromate separation of barium, strontium and lead

	Before Mg addition		After Mg addition		K_S for $MCrO_4$
	$[M^{2+}], M$	$[M^{2+}][CrO_4^{2-}]$	$[M^{2+}], M$	$[M^{2+}][CrO_4^{2-}]$	
Ba^{2+}	1.2×10^{-9}	2.4×10^{-10}	3×10^{-3}	4×10^{-4}	2×10^{-10}
Sr^{2+}	1.6×10^{-10}	2.8×10^{-12}	8×10^{-4}	1×10^{-4}	4×10^{-5}
Pb^{2+}	8×10^{-20}	1.2×10^{-20}	5×10^{-13}	6×10^{-14}	10^{-13}

The solubility product of lead chromate is not exceeded, but that of strontium chromate is slightly exceeded. The supersaturation limit, however, would not be exceeded, and no precipitation would occur, though some co-precipitation of strontium chromate with barium chromate would be expected. The solubility products of lead and strontium sulphate (p. 102) are not exceeded.

To the solution remaining after addition of magnesium ions, zinc ions are added. Zinc forms a much more stable EDTA chelate ($K = 3 \times 10^{16}$) than magnesium, and thus is more effective for displacing lead and, especially, strontium from their EDTA chelates [10]. Unfortunately, the reaction is carried out in an ethanolic solution to facilitate precipitation of strontium chromate, for which medium the relevant equilibrium constants are not known. There is also a competitive reaction because of the ammonia present. Hence a meaningful calculation cannot be made.

3.2.3 Copper–Tin Group [11]

Copper(II), mercury(II), bismuth(III), cadmium(II), tin(II), tin(IV), antimony(III), arsenic(III), arsenic(V), [selenium(IV), selenium(VI), tellurium(IV), tellurium(VI), and molybdenum(VI)].

The Copper–Tin Group cations are precipitated as sulphides by hydrogen sulphide from the solution remaining after the separation of the Calcium Group cations. The solution must first be freed from alcohol by boiling and then, in order to achieve rapid and complete precipitation of some of these sulphides, it is essential to adjust the acidity to be within the range 0.3–0.5M (pH 0.5–0.3). When this has been done, the hydrogen sulphide is added as a solution in acetone — this has many advantages over other methods of producing sulphide ions in the solution (see p. 154). The need to control the acidity is explained below.

Hydrogen sulphide is a dibasic acid which behaves as a weak acid in aqueous solution in both its dissociation stages:

$$H_2S \rightleftharpoons H^+ + HS^- \qquad K_1 \sim 10^{-7}$$

$$HS^- \rightleftharpoons H^+ + S^{2-} \qquad K_2 \sim 10^{-13}$$

Suppose the hydrogen ion concentration is adjusted to $0.4M$. An aqueous solution saturated with hydrogen sulphide at room temperature is $0.1M$ in hydrogen sulphide [12]. Although such a concentration cannot be attained by bubbling hydrogen sulphide through an aqueous solution under the conditions existing in qualitative analysis, it can probably be achieved by adding the solution of hydrogen sulphide in acetone. The concentrations of S^{2-} and HS^- ions in the solution, therefore, neglecting the presence of acetone, may be calculated as follows:

$$K_1 = \frac{[H^+] [HS^-]}{[H_2S]} = 10^{-7}$$

Assuming $[S^{2-}]$ to be very small compared to $[H_2S]$ and $[HS^-]$,

$$[H_2S] = 0.1 - [HS^-]$$

and

$$\frac{0.4 [HS^-]}{0.1 - [HS^-]} = 10^{-7}$$

so

$$[HS^-] = 2.5 \times 10^{-8}M \ .$$

Similarly,

$$K_2 = \frac{[H^+] [S^{2-}]}{[HS^-]} = 10^{-13}$$

so

$$\frac{0.4 [S^{2-}]}{2.5 \times 10^{-8}} = 10^{-13}$$

and

$$[S^{2-}] = 6 \times 10^{-21}M$$

confirming that the S^{2-} concentration in the acidic medium is very small. The concentration in the acetone medium should be even lower. By use of this value for the sulphide ion concentration, and assuming that the metal ion concentration in the 2 ml of solution is $0.05M$, the ionic concentration product for sulphides of formula MS is found to be

$$[M^{2+}] [S^{2-}] = 0.5 \times 10^{-2} \times 6 \times 10^{-21} = 3 \times 10^{-23}$$

and for sulphides of formula M_2S_3 it is

$$[M^{3+}]^2[S^{2-}]^3 = (0.5 \times 10^{-2})^2 \times (6 \times 10^{-21})^3 = 5 \times 10^{-66}$$

Comparison of these ionic concentration products with the solubility products of the sulphides given in Table 3.5 shows which of these sulphides should precipitate under these conditions.

Table 3.5

Solubility products $(25°C)$ of some sulphides in aqueous solution [13]

Sulphide	HgS	PbS	CuS	CdS	SnS	ZnS
Solubility product (in appropriate units)	10^{-52}	10^{-27}	10^{-35}	10^{-27}†	10^{-25}	10^{-22}† 10^{-24}‡

Sulphide	MnS	CoS	NiS	FeS	Bi_2S_3	Sb_2S_3
Solubility product (in appropriate units)	10^{-10}*	10^{-20}† 10^{-25}‡	10^{-19}† 10^{-24}‡	10^{-17}	10^{-97}	10^{-93}

Values for SnS_2, MoS_3, As_2S_5, and As_2S_3 do not appear to have been measured.
*pink form　　　　　　　†on precipitation　　　　　　　‡on aging

The results reported in the literature for measurements of solubility products show considerable variations. The values given in Table 3.5 are the best available at the time of writing. Variations result partly from the change in crystal structure of some sulphides on standing, to form less-soluble sulphides [14]. For example (Table 3.5), the solubility products of zinc, nickel and cobalt(II) sulphides are significantly decreased when the precipitate has aged, owing to a change to another, more strongly bound, crystal structure; e.g. for zinc sulphide, from a hexagonal wurtzite to a cubic sphalerite structure.

According to the solubility product values in Table 3.5, under the conditions used for the calculation of the ionic concentration products for sulphides MS and M_2S_3, mercury(II), lead, copper(II), cadmium, tin(II) and bismuth sulphides would be precipitated. In addition, tin(IV), arsenic(III), antimony(III) and molybdenum(VI) sulphides precipitate, indicating that in the case of arsenic(III) and antimony(III), their solubility products must be less than the ionic concentration products (5×10^{-66}) calculated above for M_2S_3 sulphides. However, arsenic(III), antimony(III) and molybdenum(VI) exist as oxy-ions in solution, so a solubility product involving ions such as As^{3+} is of limited usefulness. Selenium and tellurium are also precipitated in this group. The precipitate is a mixture of elemental selenium or tellurium and sulphur rather than a definite compound, such as selenium sulphide, SeS.

In this scheme, tin(II), if present, is oxidized beforehand to tin(IV) by bromine water, so that tin(IV) sulphide is the species precipitated. Conversion into tin(IV) is carried out because tin(IV) sulphide dissolves much more readily than tin(II) sulphide in alkali, and thus the subsequent division of the sulphide precipitate by dissolution in alkali (p. 115) is much more satisfactory. If the alternative chloroacetic acid dissolution procedure (p. 116) is used, tin(II) sulphide dissolves readily, so the prior oxidation is then unnecessary. If tin(IV) is present, oxalic acid must be added to keep it in solution as an oxalato-complex during the adjustment of pH before addition of the hydrogen sulphide; otherwise on addition of alkali tin(IV) may precipitate as so-called metastannic acid (H_4SnO_4), which is difficult to redissolve on acidification.

Arsenic(V) can be reduced by hydrogen sulphide to arsenic(III) which precipitates as the sulphide. Precipitation occurs readily only at high acidities (ca. $8M$); the rate of reduction is otherwise too slow for practical purposes. As the other sulphides do not precipitate at such a high acidity some separation schemes precipitate arsenic in $8M$ acid and the rest of the group from $0.25M$ acid.

In the present scheme, however, any arsenic(III) will have been oxidized to arsenic(V) by the bromine water added to oxidize tin(II). In order to avoid the complication of precipitation at two acidities, the arsenic(V) is reduced to arsenic(III) by boiling with ammonium iodide solution [15], before addition of the hydrogen sulphide:

$$AsO_4^{3-} + 2H^+ + 3I^- \longrightarrow AsO_3^{3-} + I_3^- + H_2O$$

The iodide ion does not reduce tin(IV) under these conditions.

Effect of Complex Formation on Sulphide Precipitation

Cadmium sulphide may be difficult to precipitate. It is commonly stated that the reason for this is the use of too high an acidity [16]. A calculation of the acidity required to prevent precipitation shows this statement to be incorrect.

The solubility product, $[Cd^{2+}] [S^{2-}] = 10^{-27}$. As $[Cd^{2+}] = 10^{-2}M$, then for precipitation to begin, the sulphide ion concentration must be at least

$$[S^{2-}] = \frac{10^{-27}}{10^{-2}} = 10^{-25}M$$

The hydrogen ion concentration required to reduce the sulphide ion concentration to this value can be calculated from the equation

$$K_1 K_2 = \frac{[H^+]^2 [S^{2-}]}{[H_2 S]} = 10^{-20}$$

As $\qquad [H_2 S] = 0.1 M$,

$$[H^+]^2 = \frac{10^{-20} \times 0.1}{10^{-25}} = 10^4$$

and $\qquad [H^+] = 100 M$

In other words, to prevent precipitation the acidity would have to exceed $100 M$, which is impossible. In fact, precipitation is prevented by hydrochloric acid at a concentration of slightly more than $0.4 M$. This arises because cadmium forms a series of chloro-complexes, $[CdCl_n]^{(2-n)}$ ($n \leqslant 4$). The complexes are not particularly stable, but sufficiently so to influence the cadmium ion concentration markedly when the chloride ion concentration is high. It may be mentioned here that in the absence of chloride, cadmium sulphide is precipitated from $5 M$ sulphuric acid [17].

The cadmium ion concentration in the solution for cadmium sulphide precipitation, when the solution is $0.4 M$ in hydrochloric acid, can be calculated as follows. The overall stability constants for the cadmium chloride complexes [cf. Eqs. (2.52) and (2.53)] are:

$$\beta_1 = \frac{[CdCl^+]}{[Cd^{2+}][Cl^-]} = 30 \; ; \qquad \beta_2 = \frac{[CdCl_2]}{[Cd^{2+}][Cl^-]^2} = 160$$

$$\beta_3 = \frac{[CdCl_3^-]}{[Cd^{2+}][Cl^-]^3} = 250 \; ; \qquad \beta_4 = \frac{[CdCl_4^{2-}]}{[Cd^{2+}][Cl^-]^4} = 60$$

The total cadmium concentration, $[Cd]_T$, is given by

$$[Cd]_T = [Cd^{2+}] + [CdCl^+] + [CdCl_2] + [CdCl_3^-] + [CdCl_4^{2-}]$$

Substituting for the various chloro-complex concentrations, by using the expressions above, gives:

$$[Cd]_T = [Cd^{2+}] (1 + \beta_1 [Cl^-] + \beta_2 [Cl^-]^2 + \beta_3 [Cl^-]^3 + \beta_4 [Cl^-]^4)$$

In the solution for precipitation, $[Cd]_T = 0.5 \times 10^{-2}M$ and $[Cl^-] \sim 1M$ (from the hydrochloric acid added to precipitate the Silver Group and subsequently adjust the pH for the Copper–Tin Group precipitation). Therefore

$$0.5 \times 10^{-2} = [Cd^{2+}] (1 + 30 + 160 + 250 + 60)$$

$$= 500 [Cd^{2+}]$$

Thus $[Cd^{2+}] = 1 \times 10^{-5}M$, 0.2% of the total cadmium concentration.

As the sulphide ion concentration at this acidity has already been calculated to be $6 \times 10^{-21}M$, the ionic concentration product of cadmium sulphide in the solution is $6 \times 10^{-21} \times 1 \times 10^{-5} = 6 \times 10^{-26}$. This value only slightly exceeds the solubility product, so at most, only a small proportion of the cadmium will be precipitated under these conditions. Further increase in the hydrochloric acid concentration can similarly be shown to prevent precipitation completely. The presence of bromide ions (from reduction of the bromine water) and iodide ions (from the ammonium iodide) also reduces the cadmium ion concentration by complex formation, and should be taken into account in a completely rigorous calculation.

The considerations above make clear the need for careful adjustment of acidity and control of the chloride ion concentration for the precipitation of cadmium sulphide. Similar considerations apply to the precipitation of tin(II) sulphide. It is found that acidities nearer the lower end of the recommended values ($0.3M$) give best results. Mercury(II), copper(II), and bismuth(III) also form chloro-complexes, but the solubility products of their sulphides are so small that complex formation does not interfere with their sulphide precipitations. Molybdenum is precipitated as a brown sulphide, but before precipitation occurs, a blue colour (molybdenum blue), indicative of reduction of some molybdenum(VI) is observed. To ensure complete precipitation of molybdenum the solution must be boiled.

Behaviour of Molybdenum, Selenium and Tellurium

Because of oxidation by the bromine water, any selenium will be present as selenate ions which, like arsenate ions, only give a precipitate from a strongly acidic solution. Selenate ions are not reduced to selenite by iodide ions, but can be reduced to selenite ions by evaporation of the solution until fumes of sulphur trioxide appear. This concentrates the hydrochloric acid sufficiently to bring about the reduction of the selenate ions [18]. Selenite ions then react with hydrogen sulphide on dilution to 0.3–$5M$ acidity, to give a precipitate of elementary selenium and sulphur, which is initially yellow but becomes red on heating.

Tellurium, present as tellurate ions, is first reduced by the hydrogen sulphide to tellurite ions and is reduced further to give a brown precipitate of tellurium and sulphur.

Mechanism of Sulphide Precipitation

Direct interaction of the metal cations with sulphide ions, S^{2-}, is improbable because of the extremely small concentration of the sulphide ions (ca. 1 ion per ml) [13]. Reaction with HS^- ions is much more probable. According to the principles outlined on p. 83, ions such as mercury(II) should bond more strongly with HS^- ions than with hydroxide ions (OH^-). The existence of $Hg(SH)_2$ in solution has been demonstrated by Treadwell and Schaufelberger [19].

It is possible, therefore, that a precipitate of $Hg(SH)_2$ is initially formed which loses hydrogen sulphide during or after the precipitation, to form the sulphide:

$$Hg^{2+} + 2HS^- \longrightarrow Hg(SH)_2(s)$$

$$Hg(SH)_2(s) \longrightarrow H_2S + HgS(s)$$

Such a mechanism requires that the final precipitate should contain some of the released hydrogen sulphide; this is found to be so [19]. The reactions of the oxy-anions also probably occur through formation of HS^- species, but are certain to be quite complicated. It should be remembered that the mechanism does not affect the solubility products, as these depend *solely* on the initial and final states of the system [14].

Co-precipitation

The discussion above accurately describes the precipitation or non-precipitation of the Copper-Tin Group cations when only *one* of the ions is present. For mixtures, however, separations are not quite so clear-cut because the effects of co-precipitation may be significant. Co-precipitation becomes more extensive the less soluble the sulphide, and thus zinc sulphide, although not itself precipitating in the Copper–Tin Group, co-precipitates appreciably with other sulphides.

In addition to some bulk co-precipitation, post-precipitation (p. 62) of Zinc Group sulphides occurs. It begins immediately after precipitation of the main precipitate has finished, and continues for an hour or more. Thus early separation of the precipitate is beneficial to avoid contamination with (and loss of) Zinc Group metals.

The reason for such post-precipitation phenomena is not clearly understood, although it has received some detailed study [20]. It would appear, however, that the properties of the post-precipitated sulphides are influenced by the original sulphide precipitate. For example, zinc sulphide post-precipitated on mercury(II) sulphide, although ostensibly existing as a phase of pure zinc sulphide around the mercury(II) sulphide particles, does not completely dissolve in $3M$ hydrochloric acid, as zinc sulphide itself does [14]. It is possible that diffusion of zinc ions into the precipitate results in formation of a solid solution.

It has been suggested [14] that there is a higher concentration of hydrogen sulphide at the surface of the precipitate particles which promotes precipitation of the normally soluble sulphide. Such a high concentration could arise directly as a result of the mechanism of sulphide formation discussed above, or from adsorption of sulphide ions on the surface. It is also likely that once post-precipitation of, for example, zinc sulphide has occurred, transformation to the less soluble crystal structure takes place, so that redissolution is prevented. However, all normal forms of zinc sulphide are soluble in $3M$ hydrochloric acid, so such explanations are not completely satisfactory.

Parting of the Copper–Tin Group

As the name suggests, the members of the group are first separated into two sub-groups, the Copper Group and the Tin Group. The Copper Group contains mercury(II), bismuth, copper, cadmium and red selenium. The Tin Group contains the other ions (tin, antimony, arsenic, molybdenum and tellurium).

The separation has traditionally been based on the solubility of the Tin Group sulphides as thio-anions, and many different reagents have been proposed for this purpose [11]. In the present scheme $0.5M$ potassium hydroxide is used. This reagent provides a good separation and is both inexpensive and easy to handle. However, the separation of mercury, and especially of selenium, is incomplete.

A complete separation of mercury is given by a recently developed scheme based on the reaction of chloroacetate ions with the Tin Group sulphides to give thiodiglycollate ions with dissolution of the precipitate [21]. Both methods of dissolution are discussed below.

(1) Dissolution of the Tin Group Sulphides in Alkali

The Tin Group sulphides are dissolved in $0.5M$ potassium hydroxide as thio-anions or, as is more likely in some instances, as hydroxothio-anions, such as $[Sn(SH)_6]^{2-}$ and $[Sn(OH)_3(SH)_3]^{2-}$, but the exact chemistry has not been elucidated. Such a process requires the high hydroxide ion concentration provided by the alkaline solution. The formation of species such as $[Sn(OH)_3(SH)_3]^{2-}$ which contain more sulphur than the original sulphide precipitated requires the provision of additional HS^- ions. These could be acquired from the hydrogen sulphide contained within the precipitate as a result of decomposition as described above for $Hg(SH)_2$ (p. 114).

Bismuth, copper and cadmium do not form stable thio-complexes and are readily separated from arsenic(III), antimony(III), molybdenum(VI) and tin(IV). As already discussed (p. 111), tin(II) sulphide also does not readily dissolve in the alkaline solution, so tin(II) is best converted into tin(IV) before precipitation with hydrogen sulphide. Of the remaining precipitates in this Group, mercury(II), selenium and tellurium are somewhat soluble in alkaline solutions, but their solubility depends on the conditions.

The difficulty of incorporating mercury quantitatively into either the Copper or the Tin Group has been the subject of much investigation and discussion. In the presence of a high sulphide ion concentration mercury dissolves as the dithiomercurate ion, HgS_2^{2-}. The use of $0.5M$ potassium hydroxide gives a solution insufficiently alkaline to permit the formation of significant concentrations of dithiomercurate(II) ions and thus allows most of the mercury to remain in the Copper Group precipitate [22]. Arsenic, antimony, molybdenum and tin sulphides are dissolved. However, an appreciable amount of mercury dissolves in the alkaline solution if any of the other sulphides that also dissolve in the alkaline solution are present. This is particularly important when tin(IV) is present, as up to 75% of the mercury may be dissolved [23]. Other reagents, such as lithium hydroxide and 1% lithium hydroxide with 5% potassium nitrate solutions, have no advantages over potassium hydroxide in this respect [11]. The use of sodium hydroxide is avoided because sodium thioantimonites, which would be formed from antimony(III) sulphide, are only sparingly soluble. Ammonium polysulphide solutions, which have also been used for this separation, dissolve appreciable amounts of mercury(II) sulphide. Most of the tellurium-sulphur mixture dissolves in $0.5M$ potassium hydroxide, but some also appears in the Copper Group [22].

Selenium forms a yellow precipitate with hydrogen sulphide in the Copper–Tin Group, which is readily soluble in $0.5M$ potassium hydroxide. However, on heating, the yellow precipitate is converted into a red form which is insoluble in the alkaline solution. In the present scheme it is necessary to boil the alkaline solution to precipitate molybdenum sulphide and, as this produces the insoluble red form of the selenium precipitate, much of the selenium remains undissolved and is confirmed in the Copper Group.

(2) Chloroacetate Dissolution of the Tin Group Sulphides [21]

Treatment of the precipitate of the Copper–Tin Group sulphides with an alkaline chloroacetate solution completely dissolves the Tin Group sulphides [arsenic(III), antimony(III), tin(IV) and molybdenum(VI)], with the formation of the appropriate metal oxy-anions and thiodiglycollate ions; for example:

$$As_2S_3(s) + 12OH^- + 6CH_2ClCOO^- \longrightarrow$$
$$2AsO_3^{3-} + 3S(CH_2COO)_2^{2-} + 6Cl^- + 6H_2O$$

Copper(II), bismuth(III) and cadmium(II) sulphides are unaffected. Mercury(II) sulphide is also unaffected — if any has dissolved as HgS_2^{2-} in the alkaline solution it is reprecipitated on reaction with the chloroacetate ions:

$$HgS_2^{2-} + 2CH_2ClCOO^- \longrightarrow HgS(s) + S(CH_2COO)_2^{2-} + 2Cl^-$$

Thus mercury is confined to the Copper Group.

In the chloroacetate procedure most of the selenium is undissolved whereas most of the tellurium dissolves. This is similar to the behaviour observed with the alkali treatment. Selenium is found in the Copper Group and tellurium in the Tin Group.

After treatment with chloroacetate the Copper Group cations may be separated and identified as described below. However, the solution of the Tin Group oxy-anions is more simply resolved than the solution of thio-anions obtained by dissolution in alkali. The two procedures are described below.

Separation of the Copper Group Precipitate

The Copper Group precipitate may include the sulphides of mercury(II), bismuth, copper and cadmium, sulphur, the red form of the sulphur–selenium mixture, and some of the tellurium–sulphur mixture. Treatment with $4M$ nitric acid oxidizes and dissolves bismuth, copper and cadmium sulphides.

For example, bismuth sulphide is oxidized as follows:

$$Bi_2S_3(s) + 2NO_3^- + 8H^+ \longrightarrow 2NO(g) + 3S(s) + 2Bi^{3+} + 4H_2O$$

This overall reaction can be considered as a combination of successive reactions:

$$Bi_2S_3(s) \rightleftharpoons 2Bi^{3+} + 3S^{2-}$$

$$3S^{2-} + 6H^+ \rightleftharpoons 3H_2S \tag{3.2}$$

$$3H_2S + 2NO_3^- + 2H^+ \longrightarrow 2NO(g) + 3S(s) + 4H_2O \tag{3.3}$$

Some sulphate may also be formed. As reactions (3.2) and (3.3) are common to all reactions of metal sulphides with nitric acid, the critical factor determining whether or not the precipitate will dissolve is the solubility product of the precipitate. For bismuth sulphide and the more soluble sulphides (Cu, Cd), the solubility product (p. 110) is sufficiently large for dissolution to occur. For mercury(II) sulphide, this is not so, and dissolution is negligible [22]. Sulphur, selenium and tellurium are not oxidized by $4M$ nitric acid. If there is a black residue after treatment with $4M$ nitric acid it may be sulphur, selenium or tellurium containing co-precipitated sulphides such as copper or bismuth, and must not simply be assumed to be mercury(II) sulphide. A confirmatory test for mercury is always to be applied.

Mercury may be separated from selenium (and other possible components of the residue described above) by the dissolution of mercury(II) sulphide in concentrated hydrochloric acid containing bromine. There is a double action, the hydrochloric acid forming chloromercurate(II) complexes and the bromine oxidizing the sulphide to sulphate:

$$HgS(s) + 4Br_2 + 4Cl^- + 4H_2O \rightleftharpoons HgCl_4^{2-} + 8Br^- + SO_4^{2-} + 8H^+$$

The bromide formed may also complex the mercury(II), assisting in the dissolution. The solution should be tested for mercury, and any residue for selenium.

The nitric acid solution is treated with ammonia, which precipitates bismuth as a basic salt, whereas copper and cadmium remain in solution as ammines. Copper and cadmium can be separated in numerous ways. In the present scheme, copper in ammoniacal solution is precipitated by selective reduction to the metal by dithionite ions. Cadmium remains in solution.

$$[Cu(NH_3)_4]^{2+} + S_2O_4^{2-} + 4OH^- \longrightarrow Cu(s) + 2SO_3^{2-} + 2H_2O + 4NH_3$$

The standard electrode potentials show that copper is more readily reduced than cadmium, (E_0: Cd/$[Cd(NH_3)_4]^{2+}$ = −0.6 V, Cu/$[Cu(NH_3)_4]^{2+}$ = −0.07 V).

Appropriate confirmatory tests for bismuth, copper and cadmium are applied to the separated products.

Separation of the Tin Group Solution
The treatment of the solution resulting from the dissolution of the Tin Group depends upon whether the solution was obtained by the use of alkali or chloroacetate ions, as each method gives different species in solution.

1. *Alkaline dissolution*
The solution resulting from the use of 0.5M potassium hydroxide is acidified with 2M hydrochloric acid and hydrogen sulphide is added to reprecipitate the Tin Group sulphides, which may include those of tin(IV), antimony, arsenic and molybdenum, some mercury and also some tellurium–sulphur and selenium–sulphur mixtures. For tin(IV) sulphide, for example, the reaction may be written:

$$[Sn(SH)_6]^{2-} + 2H^+ \rightleftharpoons SnS_2(s) + 4H_2S(g)$$

In the present scheme use is made of the difference in the solubilities of tin(IV) and antimony(III) sulphides in hydrochloric acid to separate them from the other Tin Group cations and from each other. Tin is selectively extracted from the sulphide precipitate with 4M hydrochloric acid and antimony is selectively dissolved from the residue with 8M hydrochloric acid:

$$SnS_2(s) + 4H^+ + 6Cl^- \rightleftharpoons SnCl_6^{2-} + 2H_2S$$

$$Sb_2S_3(s) + 6H^+ + 8Cl^- \rightleftharpoons 2SbCl_4^- + 3H_2S$$

The residue from the hydrochloric acid extraction is dissolved by oxidation with hypochlorite ions (or hydrogen peroxide if tellurium and molybdenum are absent). Addition of ammonia and magnesia mixture to the solution precipitates magnesium tellurate, magnesium ammonium arsenate and mercury(II) aminochloride and metallic mercury, whereas the molybdate ions remain in solution and are thus separated.

The precipitate is dissolved in $4M$ hydrochloric acid, and tellurium is selectively precipitated as the element by reduction of the tellurate with iodide in a saturated sodium sulphite solution. At the same time the iodide reduces arsenate to arsenite and reacts with mercury(II) to form tetraiodomercurate, HgI_4^{2-}. Arsenic may be detected in the solution by precipitation of yellow 12-molybdo-arsenic acid, without interference from the mercury.

(2) *Chloroacetate Dissolution*

The solution of Tin Group metal oxy-anions resulting from chloroacetate treatment may be tested directly for tin, arsenic, antimony, molybdenum and tellurium with little further separation. This avoids the problems associated with incomplete reprecipitation of the sulphides on acidification of the thio-anion solution and the need for careful control of the acidity to separate antimony and tin sulphides.

3.2.4 Iron Group [24]

Iron(III), chromium(III), aluminium(III), [titanium(IV), zirconium(IV), cerium(III), uranium(VI), vanadium(IV), vanadium(V), beryllium(II), thorium(IV)].

The Iron Group cations are precipitated as hydrated oxides from ammoniacal solution.

The solution remaining after the separation of the Copper–Tin Group precipitate is boiled to expel all the hydrogen sulphide. If the hydrogen sulphide is incompletely expelled, sulphides of the metals of the Zinc Group may precipitate as the pH is increased to precipitate the Iron Group. Any iron originally present will have been partially reduced to iron(II) by the hydrogen sulphide and must be reoxidized to iron(III), because iron(II) hydroxide is not precipitated by ammonia in the presence of ammonium chloride (see below). The oxidation is accomplished by boiling with concentrated nitric acid. Hydroxide ions are added to precipitate the hydrated oxides of the Iron Group cations. These precipitates are generally referred to as hydroxides and written as $M(OH)_n$. The optimal hydroxide ion concentration is 10^{-3} - $10^{-4}M$, and is controlled by the addition of an ammonia–ammonium chloride buffer solution.

Ammonia is used as the source of hydroxide ions for the following reasons:

(i) no metal ions are introduced into the solution (unless the ammonia solution has been stored in a glass container for a long time);

(ii) the hydroxide ion concentration is easily controlled within the range 10^{-3}-$10^{-4}M$. At higher hydroxide ion concentrations, many other cations, for example nickel, cobalt, manganese and magnesium, would also precipitate, whilst aluminium and beryllium would remain in solution as soluble hydroxo-complexes;

(iii) ammonia forms soluble ammines with certain metal ions, such as nickel, cobalt and zinc, which are thus prevented from precipitating in this group.

Many organic anions, for example oxalate, citrate and tartrate, and also phosphate, fluoride, silicate and borate ions, interfere (by precipitation or complexation) with the complete separation of the Iron Group. The mode of interference and the methods used to eliminate these ions and so prevent interference are dealt with on p. 127, because these anions may affect the separation of more than one group of metal ions.

The solubility products of the relevant hydrated metal oxides (taken as being hydroxides), are given in Table 3.6.

Table 3.6
Solubility products ($25°C$) of some 'hydroxides'

Hydroxide	$VO(OH)_2$	$Be(OH)_2$	$UO_2(OH)_2$	$Zn(OH)_2$	$Co(OH)_2$	$Ni(OH)_2$
Solubility product (in appropriate molar units)	10^{-24}	10^{-21}	10^{-21}	10^{-17}	10^{-15}	10^{-15}

Hydroxide		$Fe(OH)_2$	$Mn(OH)_2$	$Mg(OH)_2$	$Fe(OH)_3$	$Al(OH)_3$
Solubility product (in appropriate molar units)		10^{-15}	10^{-13}	10^{-11}	10^{-38}	10^{-32}

Hydroxide		$Cr(OH)_3$	$Ce(OH)_3$	$Zr(OH)_4$	$Th(OH)_4$	$Ti(OH)_4$
Solubility product (in appropriate molar units)		10^{-30}	10^{-22}	10^{-48}	10^{-45}	10^{-43}

The pH of the ammonia–ammonium chloride buffer solution in which the Iron Group metals are precipitated is between 10 and 11, that is, the hydroxide ion concentration is between 10^{-3} and $10^{-4}M$. Assuming it to be the lower value, $10^{-4}M$, and assuming the metal ion concentration to be $10^{-2}M$, the ionic concentration products in the solution are:

for 'hydroxides' $M(OH)_2$, $10^{-2} \times (10^{-4})^2 = 10^{-10}$

for 'hydroxides' $M(OH)_3$, $10^{-2} \times (10^{-4})^3 = 10^{-14}$

for 'hydroxides' $M(OH)_4$, $10^{-2} \times (10^{-4})^4 = 10^{-18}$

From these calculations all the hydroxides listed in Table 3.6 would be expected to precipitate. In fact, all the metal(III) and metal(IV) ions precipitate but of the metal(II) ions only beryllium precipitates. Thus vanadium(IV), beryllium(II), uranium(VI), iron(III), chromium(III), cerium(III), zirconium(IV), thorium(IV) and titanium(IV) precipitate whereas zinc(II), cobalt(II), nickel(II), iron(II), manganese(II) and magnesium(II) do not. The reason for the lack of precipitation of the last six ions is sometimes stated to be the result of the buffering action of ammonium chloride [25]. But the calculations carried out above take into account this buffering action and still predict that these metals should precipitate. Thus other factors must be responsible for preventing precipitation of these cations.

As in other separations within Groups, complex formation plays an important role in controlling the metal ion concentration. In the Iron Group separation, the concentration of ammonia is high, about $3M$, and the formation of ammines is an important factor. Stability constants for some relevant ammines are given in Table 3.7. The use of the available equilibrium data is of little value for the metal(III) ammines because of the slow reactions of these complexes (p. 52).

Chromium(III) forms a series of ammines which can be hydrolysed by boiling the solution. If a large excess of ammonia has been added to precipitate the Iron Group, appreciable quantities of violet chromium(III) ammine will be formed. Such solutions require prolonged boiling, or acidification and more careful reprecipitation with less ammonia, to precipitate the chromium. Boiling, or prolonged contact with air, however, may cause the precipitation of manganese in this Group. Although manganese(II) hydroxide is precipitated only very slightly or not at all, aerial oxidation causes the formation of insoluble hydrated manganese(IV) oxides. Stability constants for metal(IV) ammines are not available, but are likely to be very small.

The effect of ammine formation on the ionic concentration product of M^{2+} metal ions may be calculated as follows. Assuming that the ammines given in Table 3.7 are the most important, and that other complexes do not significantly influence the metal ion concentration, the total concentration of metal ion species, $[M]_t$, is

$$[M]_t = [M^{2+}] + [M(NH_3)_n^{2+}] = 10^{-2}M$$

The formation constant, β_n, is

$$\beta_n = \frac{[M(NH_3)_n^{2+}]}{[M^{2+}][NH_3]^n}$$

By substituting and simplifying,

$$[M^{2+}] = \frac{[M]_t}{\beta_n[NH_3]^n + 1} \tag{3.4}$$

Table 3.7 gives the stability constants for six of these metal ammines. By substituting these values in Eq. (3.4) together with values for the minimal likely concentration of ammonia ($3M$) and the total concentration of the metal ion species, $[M]_t \cong 10^{-2}M$, the metal ion concentrations in the presence of ammonia can be calculated and hence the ionic concentration products. The results of these calculations are given in Table 3.7.

Table 3.7
Stability constants of some ammines (approximate)

Ammine	$Fe(NH_3)_4^{2+}$	$Co(NH_3)_6^{2+}$	$Ni(NH_3)_6^{2+}$	$Mn(NH_3)_2^{2+}$	$Zn(NH_3)_4^{2+}$	$Mg(NH_3)_2^{2+}$
Stability constant (appropriate molar units)	10^4	10^5†	10^9†	10	10^9†	10
$[M^{2+}]M$	10^{-8}	10^{-10}	10^{-14}	10^{-4}	10^{-12}	10^{-4}
Ionic concentration product $[M^{2+}][OH^-]^2$	10^{-16}	10^{-18}	10^{-22}	10^{-12}	10^{-20}	10^{-12}

†There are significant contributions from the lower ammine complexes, but these have been allowed for in the approximation used for β_n.

These results show that the ionic concentration product is significantly less than the solubility product except for magnesium and manganese. Precipitation of iron(II), cobalt(II), nickel(II) and zinc(II), therefore, should not occur. For magnesium and manganese the ionic concentration products slightly exceed the solubility products and precipitation of these metals is possible, and manganese will always be divided between this Group and the Zinc Group; only a small fraction of it appears in the Iron Group, however. Cobalt can be extensively co-precipitated (up to 75% [26]) if the conditions are not correct.

Vanadium is almost all present as vanadium(IV) after the Copper–Tin Group separation, whether originally present in oxidation state (IV) or (V), but is mostly oxidized to vanadium(V) by boiling with nitric acid before the Iron Group separation. Precipitation of vanadium(V) and vanadium(IV) is incomplete in the ammonia–ammonium chloride solution, however, probably because of the complicated polymerization reactions that occur, but if an excess of iron(III) is added, the vanadium(V) is quantitatively precipitated as iron(III) vanadate. The presence of vanadium will usually have been suspected before this stage because of the formation of blue colours, owing to reduction to vanadium(IV), after, for example, treatment with hydrogen sulphide.

Separation of the Iron Group Precipitate
Treatment of the Iron Group precipitate with sodium hydroxide solution

dissolves hydrated aluminium and beryllium oxides and iron(III) vanadate, to give tetrahydroxoaluminate, tetrahydroxoberyllate and vanadate ions, respectively. For example:

$$Al(OH)_3 + OH^- \rightleftharpoons Al(OH)_4^-$$

On dilution and boiling, however, beryllium is reprecipitated as the hydrated oxide, because the decreased hydroxide ion concentration is insufficient to prevent precipitation:

$$Be(OH)_4^{2-} \rightleftharpoons Be(OH)_2(s) + 2OH^-$$

Thus only aluminium and vanadium remain in solution. Aluminium is readily separated from vanadium by neutralizing with hydrochloric acid. The tetrahydroxoaluminate is converted into a precipitate of hydrated aluminium oxide, but vanadate ions remain in solution. The precipitate is tested for aluminium and the solution for vanadium.

The residue insoluble in the sodium hydroxide solution, *in the absence of less common metals in the test sample,* contains only hydrated iron(III) and chromium(III) oxides. Hydrated zirconium oxide will also be present if zirconium nitrate solution has been added to remove phosphate ions (p. 127). The residue is heated with hydrogen peroxide and alkali, which converts the chromium(III) oxide into soluble chromate ions:

$$Cr_2O_3.aq(s) + 3H_2O_2 + 4OH^- \longrightarrow 2CrO_4^{2-} + 5H_2O$$

but does not affect the iron(III) or zirconium oxides. The solution is tested for chromium and the precipitate for iron. Any manganese that may inadvertently be precipitated in this Group will follow iron through the separation, and may also be tested for in this final residue. The presence of zirconium does not affect these tests.

If less common metals are present, the residue which is insoluble in dilute sodium hydroxide solution may contain the hydrated oxides of titanium(IV), zirconium(IV), uranium(VI), cerium(III), beryllium(II), thorium(IV), chromium(III), iron(III) and also ammonium diuranate. After dissolution of these compounds in dilute acid, titanium and zirconium can be separated from the other ions by precipitation as their phosphates.

Treatment of the phosphate precipitate with dilute sulphuric acid and hydrogen peroxide converts titanium(IV) phosphate into a soluble yellow peroxotitanate, which is a sufficient confirmatory test for titanium. To test for zirconium, the zirconium phosphate is rendered soluble by conversion into an anionic oxalato-complex by extraction with a solution of ammonium oxalate and oxalic acid (3:1 molar ratio). Addition of sodium hydroxide solution precipitates hydrated zirconium oxide which is easily soluble in concentrated hydrochloric acid to give a solution for the confirmatory test for zirconium with mandelic acid. Under the conditions of the zirconium test, only zirconium and hafnium give precipitates with this reagent.

If zirconium has previously been added to remove phosphate ions, the presence of zirconium in the test sample must be checked by testing the original sample, in the absence of added zirconium.

After the removal of the titanium and/or zirconium phosphate precipitate, the excess of phosphate ions in the solution must be removed, otherwise undesired precipitates of chromium(III) and uranium(VI) may be obtained when the solution is treated with boiling sodium hydroxide solution during the next separation step. Removal is conveniently carried out by careful precipitation with zirconium. The phosphate-free solution is treated by boiling with hydrogen peroxide and sodium hydroxide solution to convert uranium(VI) into soluble peroxouranate(VI) ions and chromium(III) into soluble chromate ions. Other ions precipitate as hydrated oxides. Uranium is separated from chromium by the precipitation of uranium(VI) phosphate from an acetic acid medium. The precipitate is tested for the presence of uranium and the solution for chromium.

Beryllium is separated from the hydrated oxides which precipitate on treatment with sodium hydroxide and hydrogen peroxide solution by dissolution as tetrahydroxoberyllate ions, $Be(OH)_4^{2-}$, in *cold* 2*M* sodium hydroxide solution, and is tested for in the solution. The residue, which may contain cerium(IV), thorium(IV) and iron(III) hydrated oxides, is dissolved in 4*M* sulphuric acid. Oxalic acid is added, which reduces cerium(IV) to cerium(III), and precipitates cerium(III) and thorium(IV) oxalates. Iron(III) and any manganese (which has inadvertently been precipitated in this Group) remain in solution as oxalato-complexes, and iron is tested for in this solution. Addition of saturated ammonium oxalate solution to the precipitate greatly increases the oxalate ion concentration and thorium oxalate slowly dissolves as trioxalato-thorate(IV) ions. Cerium(III) oxalate does not dissolve in the ammonium oxalate solution, and cerium is tested for in the precipitate. Acidification of the thorium solution reduces the oxalate ion concentration, and reprecipitates thorium oxalate. This reaction is sufficient to confirm the presence of thorium.

3.2.5 Zinc Group

Zinc(II), manganese(II), cobalt(II), nickel(II)

The cations are precipitated as their sulphides by the addition of hydrogen sulphide in acetone to the ammoniacal solution remaining after removal of the Iron Group precipitates. In the alkaline solution, pH 9–10, the sulphide ion concentration is much higher than in the acidic solution used for the precipitation of the Copper–Tin Group. By means of the equation on p. 109, the sulphide ion concentrations in solutions of pH 9 and 10 can be calculated as follows.

The total concentration of hydrogen sulphide species, $[H_2S]_T$, is given by

$$[H_2S]_T = [H_2S] + [HS^-] + [S^{2-}] \cong 0.1M$$

From the expressions given for K_1 and K_2 for H_2S on p. 109,

$$[H_2S]_T = \frac{[H^+]^2 [S^{2-}]}{K_1 K_2} + \frac{[H^+] [S^{2-}]}{K_2} + [S^{2-}] = 0.1M$$

At pH 9 and 10, $[H^+] = 10^{-9}M$ and $10^{-10}M$, respectively; thus substituting these values, and those for K_1 and K_2 into the equation above gives:

$$pH\,9\;:\left(\frac{10^{-18}}{10^{-20}} + \frac{10^{-9}}{10^{-13}} + 1\right)\,[S^{2-}] = 0.1M,$$

therefore $[S^{2-}] = 10^{-5}M$

$$pH\,10:\left(\frac{10^{-20}}{10^{-20}} + \frac{10^{-10}}{10^{-13}} + 1\right)\,[S^{2-}] = 0.1M$$

therefore $[S^{2-}] = 10^{-4}M$

The ammine complexes of the metal ions have also to be taken into account, however, and the values given for the free metal ion concentrations in Table 3.7 can be used. Thus, the ionic concentration products for the sulphides are

	ZnS	MnS	CoS	NiS
pH 9	10^{-17}	10^{-9}	10^{-15}	10^{-19}
pH 10	10^{-16}	10^{-8}	10^{-14}	10^{-18}

These values are all in excess of the solubility products of zinc(II), cobalt(II), nickel(II) (after aging) and manganese(II) sulphides (Table 3.5), so these sulphides all precipitate readily.

With the most soluble of these sulphides, manganese(II) sulphide, the concentration of manganese(II) ions in the solution after precipitation is at most $10^{-3}M$, as is shown by simple calculation [K_S for manganese(II) sulphide $\cong 10^{-10}$]. Nickel(II) sulphide has a strong tendency to form a dark brown colloidal solution. This can be coagulated by making the solution weakly acidic with acetic acid and boiling.

Cadmium sulphide will precipitate in this Group if the conditions for precipitation in the Copper–Tin Group have not been strictly adhered to. Thallium also precipitates, as thallium(I) sulphide, if it was incompletely precipitated in the Silver Group. The solubility product of thallium(I) sulphide is 10^{-20}. It can be shown, as on p. 110, that thallium(I) sulphide would not precipitate in the Copper-Tin Group, because its ionic concentration product would be unlikely to exceed 10^{-25}. Under the conditions for the precipitation of the Zinc Group, however, its ionic concentration product would be about 10^{-11}, so that it would be completely precipitated with the Zinc Group sulphides. Also, if all the iron(II) present has not been oxidized before precipitation of the Iron Group, black iron(II) sulphide ($K_S \cong 10^{-17}$) will also precipitate in this Group.

Separation of the Zinc Group Sulphide Precipitate

The precipitated Zinc Group sulphides are dissolved with *aqua regia*. The sulphide is oxidized to sulphate ions under these conditions. The resulting solution is evaporated to dryness, and the residue is dissolved in $1M$ acetic acid. Addition of hydrogen sulphide to this weakly acidic solution at about $100°$ precipitates only zinc sulphide [and any residual cadmium and thallium(I) as sulphides]. The precipitate is tested for zinc, but thallium, which interferes in the test, must first be removed by precipitation with iodide ions.

The solution from which zinc(II) sulphide has been separated is boiled to expel hydrogen sulphide and made ammoniacal. Manganese(II) is oxidized by bromine water to manganese(IV), which precipitates as hydrated manganese(IV) oxide. Any iron(II) carried over into the Zinc Group is oxidized to iron(III), which also precipitates as the hydrated oxide. It does not interfere in the confirmatory test for manganese, however, which is carried out by oxidation of the precipitate with sodium bismuthate. Purple permanganate ions are formed if manganese is present.

Nickel and cobalt remain in solution as nickel(II) and cobalt(III) ammines. Further separation is not necessary, and they are tested for individually in the solution.

3.2.6 Magnesium Group

Magnesium, lithium, sodium (potassium, ammonium)

This group comprises only those cations which have not been removed in previous groups. As there is no reagent which will precipitate them as a group, the separation within the group is carried out on the solution obtained after removal of the Zinc Group precipitate.

Separation within the Magnesium Group

Magnesium ions are separated by forming the magnesium 8-hydroxyquinoline complex and extracting this into chloroform from $1M$ ammonia solution. After evaporation of the chloroform, the complex is destroyed by strong heating to give magnesium oxide, organic decomposition products and 8-hydroxyquinoline [27]. The oxide is dissolved in dilute hydrochloric acid and a confirmatory test for magnesium is applied. After the extraction of magnesium, the aqueous layer is tested for sodium and lithium. Potassium and ammonium are tested for during the preparation of the solution for anion analysis (p. 171).

3.2.7 Anion Interferences in Group Separation of Cations

Certain anions interfere with the separation of cations into the Groups. The interference arises either from complexation which prevents precipitation of a metal in the correct Group, or from the formation of undesirable precipitates. These anions include thiosulphate, silicate, phosphate, fluoride, borate and organic anions such as benzoate, oxalate, tartrate and citrate. The interference caused by each anion is considered in turn.

Thiosulphate ions interfere by complexing with silver(I) ions, which prevents the formation of silver chloride. Certain metal thiosulphates are converted into sulphides, which precipitate in the Silver Group. In the presence of acid and of nitrate ions, thiosulphate ions may be converted into sulphate ions and hence Calcium Group metals may precipitate in the Silver Group as sulphates. Thiosulphate ions are conveniently removed by acidification before beginning the systematic cation scheme [28].

When *silicate* ions are present, some hydrated silica is precipitated in the Silver Group, but it does not interfere with the confirmatory tests. The remaining silicate precipitates in the Iron Group and causes difficulty with the confirmation of aluminium, but is readily removed immediately after precipitation of the Copper-Tin Group. Evaporating the solution just to dryness with hydrochloric acid two or three times converts the hydrated silica into a granular form which can readily be separated by centrifuging. The residue from the evaporation should not be baked too strongly because iron(III), aluminium(III), chromium(III) and magnesium(II) may be converted into insoluble forms that will redissolve only on fusing with alkali.

Phosphate ions precipitate the Iron and Zinc Group ions and magnesium ions in the Iron Group because they form sparingly soluble phosphates under the alkaline conditions. To prevent this, phosphate ions are removed by precipitation as zirconium phosphate immediately after separation of the Copper-Tin Group. It is advantageous to delay the identification and separation of phosphate until arsenic has been removed because, otherwise, if the test for phosphate is done without due care, arsenate ions give a yellow precipitate of ammonium 12-molybdoarsenate which can be confused with yellow ammonium 12-molybdophosphate. The excess of zirconium ions precipitates in the Iron Group.

Fluoride ions interfere by precipitating magnesium, cobalt, nickel and zinc fluorides when the solution is made alkaline to precipitate the Iron Group and by forming soluble complexes with Iron Group metals, for example FeF_6^{3-} and AlF_6^{3-}, and thus preventing their precipitation. Fluoride ions may be removed as volatile hydrogen fluoride after the separation of the Copper-Tin Group.

This is done by evaporating the solution to near dryness two or three times with concentrated hydrochloric acid.

Borate ions interfere in ammoniacal solutions because many transition metal borates are only sparingly soluble under these conditions and may precipitate or co-precipitate on addition of alkali [29]. Borate ions are removed by volatilization as methyl borate after separation of the Copper–Tin Group.

Some *organic anions* interfere either by forming precipitates in the Iron Group or by forming soluble complexes. Citrate and tartrate ions, for example, form strong complexes with iron(III), aluminium(III) and chromium(III) and prevent precipitation of these cations. Many benzoates and oxalates are not soluble under alkaline conditions, and precipitate. Organic anions should be destroyed after the removal of the Copper–Tin Group, by oxidation with a mixture of sulphuric and nitric acids. As pointed out on p. 78, citrate and tartrate can act as bridging ligands in kinetically inert complexes.

From this brief discussion it is to be noted that interference, other than that of thiosulphate ions, occurs mainly when the solution is made ammoniacal to precipitate the Iron Group. All the interfering anions, apart from thiosulphate ions, are therefore removed immediately after the precipitation of the Copper–Tin Group metals. It is essential to know which of these anions are present, so that appropriate measures may be taken to remove them.

3.3 SYSTEMATIC SCHEME FOR THE SEPARATION OF ANIONS

Several attempts to develop schemes for the systematic separation and qualitative analysis of anions have been reported [30–38]. Generally, they have not proved very satisfactory because they are able to handle only a limited selection of common anions in admixture and often prove tedious.

One difficulty in producing a systematic scheme is the large number of common anions which might be included; there is also a dearth of selective precipitants and extractants for these anions. The situation is complicated by the fact that some elements exist in several anionic forms. Sulphur, for example, can be encountered as sulphide, polysulphides, sulphite, sulphate, thiosulphate, thiocyanate, dithionite, dithionate, polythionates and peroxodisulphate. Also, some anions are incompatible in admixture under certain conditions, and are thus mutually destructive; for example, arsenite and bromate, or sulphite and chromate. In such instances, unless allowance is made for incompatibility, the determination of the original composition of a mixture becomes difficult.

Some systemization is necessary both to save time and to avoid the tedium of searching individually for each of a wide range of possible anions.

The MAQA scheme described in this book is based on the work of Belcher and Weisz [39, 40]. It is the most successful of those so far devised and about three dozen anions are accommodated.

In this scheme, as in others, the anions are first converted into their sodium salts by metathesis with sodium carbonate solution. Selective precipitation procedures are then used to separate the mixture of sodium salts into groups of anions.

1. Anions precipitated as silver salts from ammoniacal solution; chloride,[†] bromide, iodide, periodate, arsenite, thiocyanate, sulphide, hexacyanoferrate(II), hexacyanoferrate(III), (selenite, tellurite, tellurate).
2. Anions precipitated as silver salts from acidic solution: bromate, iodate, (selenate). During the acidification silicate and tungstate separate as their hydrated oxides and salicylate and benzoate as the free acids.
3. Anions precipitated as calcium salts from ammoniacal solution: fluoride, phosphate, arsenate, oxalate, tartrate, (tellurate).
4. Anions precipitated as barium salts from ammoniacal solution: chromate, sulphate, (vanadate, molybdate).
5. Anions remaining in solution and tested for individually: borate, chlorate, citrate, formate, perchlorate, succinate.

Many of the anions not included can be identified by the liberation of easily identifiable volatile products on treatment with acids or other reagents, or by specific tests.

Permanganate, thiosulphate and cyanide ions interfere in the present scheme, and must be destroyed by the addition of hydrogen peroxide at the outset.

Sodium Carbonate Metathesis

The first stage in the scheme, the conversion of the anions in a sample into their water-soluble sodium salts, is achieved by metathesis with a boiling concentrated sodium carbonate solution; for example:

$$CO_3^{2-} + MX(s) \rightleftharpoons X^{2-} + MCO_3(s)$$

Metal ions that would otherwise interfere in the separation scheme are thus removed by precipitation as their carbonates or basic carbonates.

The effectiveness of the metathesis depends on the relative solubilities of the metal salt, MX, and the corresponding carbonate, MCO_3. It also depends on the carbonate ion concentration. In discussing the metathesis of barium sulphate (p. 103), it was shown that all the barium sulphate is, theoretically, converted into barium carbonate, so that all the sulphate ions are released into solution as soluble sodium sulphate. For anion analysis the amounts of sample should be chosen so as to give a $0.1M$ concentration of each anion of interest in the sodium carbonate solution.

Certain materials are not significantly affected by boiling with sodium carbonate solution, although the anions they contain are included in the scheme.

†The solution used is not sufficiently ammoniacal to dissolve silver chloride.

A typical example is silver(I) chloride. The equilibrium involved in the attempted metathesis is given by:

$$2AgCl(s) + CO_3^{2-} \rightleftharpoons Ag_2CO_3(s) + 2Cl^-$$

The solubility products for silver chloride and silver carbonate are 2×10^{-10} and 8×10^{-12} respectively. Thus, at equilibrium:

$$\frac{[Ag^+]^2[Cl^-]^2}{[Ag^+]^2[CO_3^{2-}]} = \frac{[Cl^-]^2}{[CO_3^{2-}]} = \frac{(2 \times 10^{-10})^2}{8 \times 10^{-12}} = 5 \times 10^{-9}M$$

Therefore, in a $1M$ carbonate solution, $[Cl^-] = 7 \times 10^{-5}M$.

Such a chloride ion concentration is too small to be detected with certainty by the present scheme. Therefore, materials of this type are analysed by the use of a separate procedure, described in the section on the analysis of insoluble substances (p. 213). There are also substances which become more resistant to metathesis after aging, calcining (heating in air) or fusion and these may thus need to be examined separately.

Although most metal ions are precipitated by the metathesis, some remain in solution as anionic carbonato-complexes. Such metals include copper, tin, uranium and beryllium. Also thallium(I) carbonate is somewhat soluble in the sodium carbonate solution.

Many anion separation schemes involve the neutralization of the sodium carbonate extract (after removal of any residue) by acidification and subsequent neutralization with ammonia. However, certain anions or combinations of anions are destroyed by acidification. These include thiosulphate ions:

$$S_2O_3{}^{2-} + 2H^+ \longrightarrow S(s) + SO_2(g) + H_2O$$

and mixtures of iodide and iodate ions:

$$IO_3^- + 5I^- + 6H^+ \longrightarrow 3I_2 + 3H_2O$$

In the present scheme, the test solution is acidified *only* after such anions have been precipitated and removed (thiosulphate is removed with hydrogen peroxide before the metathesis).

Examination of the Sodium Carbonate Extract

The solution from the metathesis is treated with a solution of diammine-silver(I) which precipitates a group of anions as their silver salts. A further group is precipitated either as silver salts or as free acids on acidification. Subsequent separations of the remaining anions are achieved by precipitation first with calcium nitrate and then with barium nitrate from ammoniacal solution. Each group precipitate, and the remaining solution, is treated further to identify the anions present. When passing from alkaline to acidic conditions and *vice versa*, temporary precipitation of various silver salts may occur, owing to momentary neutrality. These precipitates redissolve when further acid or alkali is added.

In the following sections, only the separations into groups will be discussed in detail, because the separations within groups are relatively straightforward, or specific tests are employed without a separation.

Precipitation with the Diamminesilver(I) Reagent

Chloride, bromide, iodide, thiocyanate, sulphide, periodate, arsenite, hexacyanoferrate(II), hexacyanoferrate(III), (selenite, tellurite, tellurate)

The reagent is prepared by adding ammoniacal ammonium carbonate solution dropwise to a silver nitrate solution until the brown precipitate of silver carbonate just redissolves. This ensures an ammonia concentration slightly in excess of that required to complex all the silver as the ammine, $Ag(NH_3)_2^+$. This excess markedly reduces the free silver ion concentration and ensures that only the relatively small number of silver salts with the lowest solubility products will be precipitated. By limiting precipitation to a small number of anions the subsequent identification in the group is simplified.

The silver ion concentration, $[Ag^+]$, after the addition of the reagent to the sodium carbonate extract, may be calculated as follows. The ammonium carbonate solution used contains 25 ml of saturated 'ammonium carbonate' solution, 10 ml of $15M$ (s.g. = 0.88) ammonia solution, and water to give a total volume of 135 ml. Solid 'ammonium carbonate' is composed of about equal amounts of ammonium hydrogen carbonate, NH_4HCO_3, molecular weight 79, and ammonium carbamate, NH_2COONH_4, molecular weight 78. The average molecular weight of the 'ammonium carbonate' is therefore 78.5; its solubility in water at room temperature is about 20% w/v, so the solution contains 5 g of 'ammonium carbonate' in 135 ml.

Carbamate ions are hydrolysed in aqueous solution and, in the presence of sufficient ammonia, complete conversion into carbonate ions is achieved. Thus an ammonium ion additional to that present in the ammonium carbamate is formed:

$$NH_2CO_2^- + H_2O \longrightarrow NH_4^+ + CO_3^{2-}$$

and an average of 1.5 ammonium ions is produced from each 'molecule' of 'ammonium carbonate' in addition to those formed by the conversion of ammonia into ammonium ions by the hydrogen carbonate ions. Therefore, the total concentration of ammonia plus ammonium ions in the ammoniacal ammonium carbonate solution is:

$$[NH_3] + [NH_4^+] = \left(\frac{10}{135} \times 15\right) + \left(1.5 \times \frac{5}{78.5} \times \frac{10^3}{135}\right)$$

$$= 1.1 + 0.7 = 1.8M$$

When this solution is mixed with the strongly alkaline sodium carbonate solution, the large excess of carbonate ions results in the almost complete conversion of the ammonium ions into ammonia:

$$NH_4^+ + CO_3^{2-} \rightleftharpoons NH_3 + HCO_3^-$$

so $[NH_3] \cong 1.8M$. Silver ions added to the resulting solution are completely converted into the ammine complex (see p. 50).

In the analysis scheme, 0.1 ml of $3.3M$ silver nitrate is mixed with 4 drops (0.2 ml) of the ammoniacal ammonium carbonate solution and 1 ml of the sodium carbonate solution, giving a total volume of 1.3 ml. The silver ammine concentration in this solution is given by

$$[Ag(NH_3)_2^+] = 3.3 \times \frac{0.1}{1.3} - [Ag^+]$$

$$= 0.26 - [Ag^+]$$

Assuming that nearly all the silver ions are complexed (see p. 50), $[Ag^+]$ can be neglected in comparison with the ammine concentration, so

$$[Ag(NH_3)_2^+] = 0.26M$$

The ammonia concentration is the same as the total original ammonium ion and ammonia concentrations (because all the ammonium ions have been converted into ammonia by the sodium carbonate), appropriately adjusted for dilution, less that complexed by the silver ions:

$$[NH_3] = 1.8 \times \frac{0.2}{1.3} - 0.26$$

$$= 0.28 - 0.26$$

$$= 0.02M$$

Thus, there is only a slight excess of ammonia over that required to complex all of the silver. A calculation similar to that on p. 50 shows the silver ion concentration to be:

$$[Ag^+] = \frac{[Ag(NH_3)_2^+]}{\beta_2[NH_3]^2} = \frac{0.26}{2 \times 10^7 \times (0.02)^2} = 3 \times 10^{-5}M$$

Similarly, if five drops of ammoniacal ammonium carbonate solution are used, the free silver ion concentration can be shown to be $2 \times 10^{-6}M$; if six drops are used, it is $5 \times 10^{-7}M$.

The anions that are precipitated by addition of the sodium carbonate extract to the silver ammine reagent are those silver salts with very small solubility products (Table 3.8), namely, hexacyanoferrate(II), hexacyanoferrate(III), sulphide, periodate, arsenite, chloride, bromide, iodide, thiocyanate, selenite, tellurite and tellurate ions. This is shown for those compounds with known solubility products in Table 3.8. As the anion concentration in the sodium carbonate solution is ca. $0.1M$, all those salts listed on the left-hand side of the table, except arsenite, will precipitate. None of those listed on the right-hand side of the table should precipitate, except for silver iodate, arsenate and phosphate. These are border-line cases, which do not precipitate in practice, probably because of supersaturation and ionic strength effects. Tellurate ions are only partially precipitated.

Table 3.8

Solubility products of some silver salts, and the minimal anion concentrations required to exceed them in alkaline medium[†]

	Precipitated			Not precipitated	
Salt	K_S (appropriate molar units)	Minimal anion conc.[‡] (M)	Salt	K_S	Minimal anion conc.[‡] (M)
AgCl	2×10^{-10}	1×10^{-4}	AgF	1	10^6
AgBr	8×10^{-13}	4×10^{-7}	AgBrO$_3$	5×10^{-5}	25
AgI	1×10^{-16}	5×10^{-11}	AgIO$_3$	3×10^{-8}	2×10^{-2}
AgSCN	1×10^{-12}	5×10^{-7}	Ag$_2$SO$_4$	2×10^{-5}	5×10^6
Ag$_2$S	10^{-50}	10^{-39}	Ag$_2$SeO$_4$	10^{-9}	250
Ag$_3$AsO$_3$	10^{-17}	1	Ag$_2$CO$_3$	10^{-11}	1.5
Ag$_2$SeO$_3$	10^{-15}	3×10^{-4}	Ag$_2$WO$_4$	10^{-11}	2.5
Ag$_4$[Fe(CN)$_6$]	10^{-41}	10^{-18}	Ag$_2$MoO$_4$	10^{-11}	2.5
			Ag$_2$CrO$_4$	10^{-12}	0.3
			Ag$_2$C$_2$O$_4$	10^{-11}	2.5
			Ag$_3$PO$_4$	10^{-20}	10^{-3}
			Ag$_3$AsO$_4$	10^{-22}	10^{-5}
			Ag$_2$HVO$_4$	10^{-14}	3×10^3

[†] Using 5 drops of silver ammine reagent, so $[Ag^+] = 2 \times 10^{-6}M$.

[‡] Calculated from $[X^{n-}] = \dfrac{K_S}{[Ag^+]^n}$, neglecting supersaturation effects.

Acidification of the Silver Ammine–Sodium Carbonate Solution
Bromate, iodate, silicate, benzoate, salicylate, (selenate, tungstate)

Acidification of the silver ammine–sodium carbonate solution, after removal of any precipitate, decomposes the ammine complex, so that the silver ion concentration is increased to $0.2M$ (allowing for some dilution of the original

0.26M silver ammine solution on acidification). This results in the precipitation of silver bromate, iodate and selenate. Because these anions are derived from fairly strong acids, they are therefore not sufficiently protonated (under the acidic conditions used) for the anion concentration to be too low for the solubility products of their silver salts to be exceeded. For example, K_1 for iodic acid is 0.18. If the initial iodate ion concentration before acidification was 0.1M, the resulting iodate ion concentration in 1.5M nitric acid, neglecting the dilution caused by acidification, is given by

$$[IO_3^-] = 0.1 - [HIO_3]$$

As
$$K_1 = \frac{[H^+][IO_3^-]}{[HIO_3]} = 0.18$$

$$[IO_3^-] = 0.1 - \frac{[H^+][IO_3^-]}{0.18} = 0.1 - \frac{1.5[IO_3^-]}{0.18}$$

therefore, $[IO_3^-] = 10^{-2}M$; that is, 10% of the iodate ions will be unprotonated.

Thus the ionic product $[Ag^+][IO_3^-]$ in the solution is $0.2 \times 10^{-2} = 2 \times 10^{-3}$, which is greatly in excess of the solubility product of silver iodate (3×10^{-8}) given in Table 3.8. Similar calculations show that silver bromate (bromic acid is a strong acid) and silver selenate (K_1 for selenic acid = 80) are also precipitated almost completely under these conditions. Furthermore, the solubility products of silver chromate and oxalate are also somewhat exceeded under these conditions. In fact, silver oxalate and chromate will precipitate if insufficient acid is added or if too great a concentration of the anion is present. However, if the experimental instructions are correctly followed, silver chromate and oxalate are not precipitated. The other anions which might be present are derived from weak acids, so they are extensively protonated under these conditions and their silver salts are not precipitated. These anions are molybdate, phosphate, arsenate, tellurate and vanadate, and the organic anions included in the scheme.

Acidification at this stage converts benzoate and salicylate ions into the free acids, which are only sparingly soluble, and therefore precipitate. Silicate and tungstate ions are also protonated, and precipitate as their hydrated oxides. The organic acids may be extracted into benzene, and are thus separated from the silver salts and the oxides.

After removal of the precipitated species, the solution is warmed to remove carbon dioxide, and then again made ammoniacal. No further silver salts are precipitated under these conditions. However, beryllium and uranium, present in the sodium carbonate solution as anionic carbonato-complexes, are precipitated from the now carbonate-free alkaline solution as beryllium hydroxide and ammonium diuranate, $(NH_4)_2U_2O_7$, respectively. Small amounts of tin, aluminium and residual silicate may also precipitate at this stage.

Precipitation with Calcium Nitrate Solution

Fluoride, phosphate, arsenate, tartrate, oxalate, (tellurate)

After the precipitation of silver salts, calcium nitrate solution is added to the remaining solution, made ammoniacal, to give a $0.1M$ calcium ion solution. This precipitates calcium fluoride, phosphate, arsenate, tartrate, oxalate and tellurate.

The formation of weak ammine and hydroxo complexes (Table 2.2) decreases the calcium ion concentration by about 20% (ammine complexes, log β_1 = 0.2, log β_2 = 0.8; hydroxo complex, log β_1 = −1.3) but this is neglected for the purpose of calculation. Table 3.9, which gives the solubility products of the relevant calcium salts (where known) and the minimal anion concentration required to exceed these solubility products under the MAQA conditions, shows that phosphate, arsenate, fluoride and oxalate should be precipitated, as the actual anion concentration should be about 5 × $10^{-2}M$. The solubility product for calcium molybdate is not known with any precision, and the value given in Table 3.9 is probably erroneous. Although the solubility product of calcium sulphate is also exceeded, the high supersaturation usually prevents any precipitation.

Table 3.9

Solubility products of some calcium salts, and the minimal anion concentrations required to exceed them[†]

Salt	K_S (appropriate molar units)	Minimal anion conc.[‡] (M)	Result
$Ca_3(PO_4)_2$	10^{-29}	10^{-13}	Precipitated
$CaHPO_4$	3×10^{-7}	3×10^{-6}	Precipitated
$Ca_3(AsO_4)_2$	10^{-18}	3×10^{-8}	Precipitated
CaF_2	3×10^{-11}	2×10^{-5}	Precipitated
CaC_2O_4	2×10^{-9}	2×10^{-8}	Precipitated
$CaMoO_4$	10^{-7}?	10^{-6}	Not precipitated
$CaSO_4$	4×10^{-5}	4×10^{-4}	Not precipitated

[†] Assuming $[Ca^{2+}] = 0.1M$.

[‡] Calculated for Ca_nX_m from $[X^{(2n/m)-}] = \left(\dfrac{K_S}{[Ca^{2+}]^n} \right)^{\frac{1}{m}}$

The precipitated calcium salts are separated into two subgroups by treatment with $1M$ acetic acid. Calcium arsenate, phosphate and tellurate dissolve, whereas the other calcium salts do not. For example, the solubility of calcium phosphate can be calculated as follows.

In $1M$ acetic acid ($K_1 = 2 \times 10^{-5}$), the hydrogen ion concentration, given by Eq. (2.19), is $4 \times 10^{-3}M$. As phosphoric acid is also a weak acid, the concentration of phosphate ions will be greatly decreased in the acidic medium. The concentration of acetic acid is much greater than that of phosphate ions, so the concentration of the latter may be calculated as follows:

$$K_1 K_2 K_3 = \frac{[H^+]^3 [PO_4^{3-}]}{[H_3PO_4]} = 10^{-21}$$

therefore $$\frac{[PO_4^{3-}]}{[H_3PO_4]} = \frac{10^{-21}}{(4 \times 10^{-3})^3} = 2 \times 10^{-14}$$

that is, most of the phosphate ions will be converted into phosphoric acid.[†]

If it is assumed that the precipitate contains 10^{-4} mole of phosphate ions (from 1 ml of $0.1M$ original sodium carbonate solution), then under the acidic conditions, if all the precipitate dissolves in the 0.5 ml of acetic acid, the phosphoric acid concentration will be $0.2M$, and thus

$$[PO_4^{3-}] = 2 \times 10^{-14} \times 0.2 = 4 \times 10^{-15}M$$

Since $3Ca^{2+} + 2H_3PO_4 \longrightarrow Ca_3(PO_4)_2(s) + 6H^+$, the calcium ion concentration will be $1.5[H_3PO_4] = 0.3M$. Hence the ionic concentration product, $[Ca^{2+}]^3 [PO_4^{3-}]^2 = (0.3)^3 \times (4 \times 10^{-15})^2 = 4 \times 10^{-31}$.

As this is less than the solubility product for calcium phosphate (10^{-29}, see Table 3.9), all the precipitate must dissolve in the acetic acid. A similar calculation shows that calcium arsenate is also completely soluble under these conditions $[K_1 K_2 K_3 (H_3AsO_4) = 10^{-20}]$, but that calcium oxalate is almost insoluble $[K_1 K_2 (\text{oxalic acid}) = 2 \times 10^{-5}]$.

Precipitation with Barium Nitrate Solution
Chromate, sulphate, (vanadate, molybdate)
     The addition of barium nitrate solution to the solution remaining after treatment with calcium nitrate solution gives a solution $0.02M$ in barium ions, which will precipitate sulphate, chromate, vanadate [41] and molybdate as barium salts. Ammine and hydroxo-complex formation (Table 2.2) has a negligible effect on the barium ion concentration. The concentration of each anion in the solution should be about $0.05M$, so the ionic concentration products for barium chromate and sulphate will be 1×10^{-3}. This is greatly in excess of the solubility products of barium chromate ($K_S = 2 \times 10^{-10}$) and barium sulphate

[†] Strictly speaking, $H_2PO_4^-$ and HPO_4^{2-} will also be formed, and the major species present will be $H_2PO_4^-$, but the calculation of $[PO_4^{3-}]$ is still valid.

($K_S = 1 \times 10^{-10}$), therefore the salts are precipitated. The solubility products of barium vanadate and molybdate are not known, but these salts are also precipitated on addition of barium nitrate.

Anions not Precipitated in the Scheme
Borate, chlorate, citrate, formate, perchlorate, succinate
    These anions are tested for individually in the solution remaining after the removal of the barium salts.

3.4 NON-SYSTEMATIC TESTS FOR ANIONS

    The scheme described above accommodates 35 anions, and includes most of the common anions. However, there are a number of common anions which are not detected by the scheme, and these have to be tested for separately. Such anions include acetate, carbonate, hydrogen carbonate, cyanate, sulphite, cyanide, thiosulphate, permanganate, nitrate, nitrite and peroxodisulphate. Additionally, special sequences of tests are included for detecting certain mixtures of anions. The chemistry of most of these is straightforward. The identification of carbonate, hydrogen carbonate and cyanate ions, however, is somewhat more complicated, and the basic chemistry is described in detail below.

Tests to Distinguish Carbonate, Hydrogen Carbonate and Cyanate in Admixture
    Tests to differentiate carbonate and hydrogen carbonate, based on precipitation with magnesium or calcium, have been described in many textbooks but have been shown to be unreliable, as have tests with mercury(II) chloride. There are some confusing discrepancies in the literature regarding the nature and colour of the precipitates formed with these anions [42, 43].
    The tests described below avoid precipitation with magnesium, and provide simple, unequivocal procedures for identifying the three ions in the presence of many other ions [42].
    The principle of the scheme is based on the following considerations.
    (a) All hydrogen carbonates are soluble in water, hence a water extraction removes all the hydrogen carbonate together with any soluble cyanate or carbonate.
    (b) Any residue may consist of insoluble carbonates and cyanates. If carbon dioxide is evolved when the residue is treated with $4M$ hydrochloric acid *insoluble carbonate and/or cyanate* must be present; the latter can be confirmed by making the acid solution alkaline and testing for ammonia.

$$OCN^- + 2H^+ + H_2O \longrightarrow NH_4^+ + CO_2(g)$$

If carbon dioxide is evolved when the residue is treated with $4M$ acetic acid in the cold, *insoluble carbonates* are present; cyanates do not react under these conditions.

The original aqueous extract is divided into three parts.

(i) One portion is treated with 4M hydrochloric acid. The evolution of carbon dioxide indicates *hydrogen carbonates and/or soluble carbonates and cyanates*.

(ii) A further portion is treated with 1 drop of phenolphthalein. A red colour indicates *soluble carbonate*, which makes the solution alkaline by hydrolysis:

$$CO_3^{2-} + H_2O \rightleftharpoons HCO_3^- + OH^-$$

Neither hydrogen carbonate nor cyanate gives this reaction. Saturated barium chloride is now added. Precipitation of barium carbonate removes carbonate ions from the solution, causing the solution to become less alkaline, and the colour to be discharged:

$$Ba^{2+} + CO_3^{2-} \rightleftharpoons BaCO_3(s)$$

$$HCO_3^- \rightleftharpoons H^+ + CO_3^{2-}$$

A slight excess of barium chloride is added, so that if hydrogen carbonate ions are present more barium carbonate is precipitated on very gently warming:

$$2HCO_3^- \rightleftharpoons CO_3^{2-} + CO_2(g) + H_2O$$

Thus the slow evolution of carbon dioxide denotes *hydrogen carbonate*.

(iii) Phenolphthalein is added to the third portion and the solution is made alkaline with sodium hydroxide solution and boiled. If ammonia is liberated, the original mixture contained an ammonium salt and this must be removed completely by boiling. The solution is then acidified with hydrochloric acid. The evolution of carbon dioxide at this stage indicates carbonate or cyanate. *Cyanate* is confirmed by adding alkali and testing for ammonia (p. 198).

When the residue is tested for cyanate [(b) above] there is no need to test for ammonium ions before the cyanate test, because the water extraction would remove all ammonium salts.

3.5 SYSTEMATIC ANALYSIS OF INSOLUBLE SUBSTANCES

The examination of substances which are insoluble in dilute or concentrated acids or *aqua regia* necessitates the use of special reagents to convert them into compounds which are soluble in water or dilute acids. The compounds may then be identified by the usual tests. In an earlier text [44] it was recommended that the portion of a sample which remained after treatment with *aqua regia* should be fused with a mixture of sodium carbonate and potassium nitrate; the fused (molten) mass was cooled, extracted with water, filtered, and the solution tested for anions such as sulphate, silicate, chromate, chloride and fluoride. The residue of metal carbonates was dissolved in hydrochloric acid and the solution examined for metals such as calcium, strontium, barium, aluminium and tin. Titanium compounds remaining were fused with potassium hydrogen sulphate, and the cooled fused material was extracted with water to give a solution of titanium(IV) sulphate. In other schemes of analysis [45] various preliminary tests and separations were made on the insoluble portion before the alkaline fusion. These included heating on charcoal, the microcosmic bead test, boiling with sodium hydroxide solution and the removal of lead and silver salts.

The present scheme of analysis [46] uses many spot tests and covers a selected number of insoluble substances:

Insoluble silver salts:	AgCl, AgBr, AgI, AgSCN
Insoluble sulphates:	$PbSO_4$, $BaSO_4$, $SrSO_4$
Ignited oxides:	Fe_2O_3, Al_2O_3, Cr_2O_3, SnO_2
Other oxides:	WO_3, MoO_3, TiO_2, ZrO_2, SiO_2
Others:	CaF_2 and other insoluble fluorides, Si, SiC, W, C, BN, $Cu_2Fe(CN)_6$, fused $PbCrO_4$, $Zr_3(PO_4)_4$, $Ti_3(PO_4)_4$, mineral silicates.

After the insoluble material has been isolated, a number of preliminary tests are carried out. These are followed by the dissolution of silver salts as thiosulphate complexes and treatment with sodium hydroxide solution. The residue is boiled with sodium carbonate solution and any undissolved material is fused with a mixture of sodium carbonate and sodium nitrate. Zirconium and titanium oxides are not readily attacked by this sequence of reactions, but may be brought into solution by fusion with potassium hydrogen sulphate.

Most of the reactions used in the separation scheme for insolubles are straightforward. However, some of the reactions of particular insoluble compounds and the general description of the chemistry underlying the separation scheme are somewhat unusual. These are given below.

Fusion Processes [47, 48]

Fusion processes are commonly used with materials insoluble in acids to give products which, though not necessarily water-soluble, are usually soluble in acids. There is a variety of types of fusion process, such as alkaline, acidic, oxidative or reductive processes, many of which are used in this scheme.

Sodium carbonate fusions act by alkaline attack of the liquid sodium carbonate flux on oxide species, for example:

$$Al_2O_3(s) + CO_3^{2-} \longrightarrow 2AlO_2^- + CO_2(g)$$

The attack may be accompanied by oxidation by oxygen from the air or by added nitrate ions, for example:

$$2Cr_2O_3(s) + 3O_2(g) + 4CO_3^{2-} \longrightarrow 4CrO_4^{2-} + 4CO_2(g)$$

Sodium peroxide is usually a more effective fusion agent or flux. Attack is by the peroxide ion, O_2^{2-}:

$$2Al_2O_3(s) + 2O_2^{2-} \longrightarrow 4AlO_2^- + O_2(g)$$

The effectiveness is such that vessels in which the fusion is carried out are also severely attacked, especially those made of nickel, iron or silver. Zirconium crucibles are reported to be the most resistant to this type of fusion. *It must also be noted that readily oxidized substances may react explosively in fusion processes.*

A comparison of the amount of dissolution of the oxides by various alkaline fluxes, for qualitative purposes, has been made [49], using as fluxes:

(a) sodium peroxide,
(b) three parts by weight of sodium carbonate and one part of sodium nitrate,
(c) equal weights of sodium carbonate and sodium nitrate,
(d) equal weights of sodium carbonate and potassium nitrate,

heated with the insoluble compounds in nickel crucibles.

It was found that *fresh* sodium peroxide is generally the most effective reagent, especially with aluminium oxide. The effectiveness of the other mixtures examined decreases in the order (b) → (d) given above.

When sodium peroxide is used, some black nickel oxide is always formed by attack on the crucible but does not interfere in the subsequent tests.

Potassium hydrogen sulphate fusion is more effective for attacking basic oxides. When potassium hydrogen sulphate is heated, dehydration occurs first, which causes appreciable foaming and sputtering, with the formation of pyrosulphate ions, $S_2O_7^{2-}$:

$$2HSO_4^- \longrightarrow S_2O_7^{2-} + H_2O(g)$$

The fusion is thus better carried out with potassium pyrosulphate (m.p. $414°$).

The active component is sulphur trioxide. A typical reaction is that of iron(III) oxide:

$$Fe_2O_3(s) + 3SO_3 \longrightarrow Fe_2(SO_4)_3$$

Fused potassium pyrosulphate has only poor oxidizing properties, so manganese(II) and chromium(III) are not oxidized, but iron(II) is readily converted into iron(III) [50]. It is useful to know that though potassium pyrosulphate is opaque at red heat during heating, it becomes transparent when the source of heat is removed, allowing ready inspection of progress of the fusion.

3.5.1 The Identification Scheme
Test for Insoluble Fluorides
    A sufficient concentration of fluoride ions is liberated from insoluble solid fluorides in the presence of acid to change the violet colour of the zirconium alizarin complex to yellow owing to the formation of colourless hexafluorozirconate(IV) ions, ZrF_6^{2-}, and yellow alizarin.

Test for Boron Nitride
    Boron is detected by means of the green flame given by volatile boron trifluoride:

$$BN(s) + 4HF \longrightarrow BF_3(g) + NH_4F$$

Nitride is detected by the liberation of ammonia from ammonium sulphate formed by heating boron nitride with concentrated sulphuric acid:

$$2BN(s) + 7H_2SO_4 \longrightarrow 2H_3BO_3 + (NH_4)_2SO_4 + 6SO_3(g)$$

Boiling with Sodium Hydroxide Solution
    Silicon, lead sulphate, fused lead chromate, molybdenum(VI) oxide, and tungsten(VI) oxide, on boiling with sodium hydroxide solution, are converted into soluble silicate, plumbate(II), chromate, molybdate and tungstate ions, respectively. For example:

$$Si(s) + 2OH^- + H_2O \longrightarrow SiO_3^{2-} + 2H_2(g)$$

$$PbSO_4(s) + 4OH^- \longrightarrow Pb(OH)_4^{2-} + SO_4^{2-}$$

$$WO_3(s) + 2OH^- \longrightarrow WO_4^{2-} + H_2O$$

Some of the zirconium and titanium phosphate is converted into the sparingly soluble hydrated oxide, with release of phosphate ions, but the reaction is too slow for conversion to be complete. Copper hexacyanoferrate(II) forms sparingly soluble hydrated copper(II) oxide and soluble hexacyanoferrate(II) and tetrahydroxocuprate(II) ions, $Cu(OH)_4^{2-}$:

$$Cu_2Fe(CN)_6(s) + 4OH^- \longrightarrow 2CuO.aq(s) + Fe(CN)_6^{4-} + 2H_2O$$

$$CuO.aq(s) + 2OH^- + H_2O \longrightarrow Cu(OH)_4^{2-}$$

After removal of undissolved material, the solution is acidified with hydrochloric acid. This precipitates hydrated silica and red copper(II) hexacyanoferrate(II). When ammonia is then added, lead hydroxide is also precipitated, as is hydrated chromium(III) oxide.[†] Tungstate and chromate ions remain in solution.

Boiling with Sodium Carbonate Solution

Of the insoluble substances remaining after treatment with sodium hydroxide solution, barium and strontium sulphates are converted into carbonates by boiling with sodium carbonate solution (see p. 103). Insoluble fluorides and silica are partially changed into the carbonate and silicate respectively:

$$CaF_2(s) + CO_3^{2-} \longrightarrow CaCO_3(s) + 2F^-$$

$$SiO_2(s) + CO_3^{2-} \longrightarrow SiO_3^{2-} + CO_2(g)$$

Silicate, fluoride and sulphate ions can be identified in the solution. The alkaline earth metal carbonates are readily dissolved by acetic acid and the metal ions identified.

Fusion with Alkali

Substances which have resisted attack by alkaline solutions may be converted into soluble compounds by fusion with alkali. These substances are aluminium oxide, chromium(III) oxide, tin(IV) oxide, titanium(IV) oxide, zirconium(IV) oxide, zirconium(IV) phosphate, titanium(IV) phosphate, boron nitride, silicon carbide, carbon, tungsten and mineral silicates. In addition, any calcium fluoride or silica which did not undergo metathesis with sodium carbonate is dissolved at this stage.

[†]Chromium(III) oxide is always present in fused lead chromate because of the thermal decomposition:

$$2PbCrO_4(s) \longrightarrow 2PbO(s) + Cr_2O_3(s) + \tfrac{3}{2}O_2(g)$$

The chromium(III) oxide partially dissolves in the sodium hydroxide solution to form chromate(III) anions, which on acidification are converted into chromium(III) ions, which precipitate when the solution is made ammoniacal.

The oxides of aluminium, chromium and tin are converted into aluminate, chromate and stannate ions, respectively, by the carbonate and peroxide reactions. Boron nitride is converted into sodium borate:

$$2BN(s) + 3CO_3^{2-} + \tfrac{5}{2}O_2(g) \longrightarrow 2BO_3^{3-} + 3CO_2(g) + 2NO(g)$$

Similarly, silicon carbide and mineral silicates form silicate ions and tungsten forms tungstate ions. For example:

$$SiC(s) + CO_3^{2-} + 2O_2(g) \longrightarrow SiO_3^{2-} + 2CO_2(g)$$

$$W(s) + CO_3^{2-} + \tfrac{3}{2}O_2(g) \longrightarrow WO_4^{2-} + CO_2(g)$$

Titanium(IV) oxide is converted into an acid-soluble form by peroxide and, to some extent, by carbonate fusion: zirconium(IV) oxide is attacked by peroxide fusion, but not by fusion with sodium carbonate. Titanium and zirconium phosphates are soluble in the fused carbonate.

After the cooled fusion mass has been boiled with water, the solution is tested for silicate, tungstate, borate and Tin and Iron Group metal ions. The residue is treated with hydrochloric acid, and the solution obtained is tested for cations, especially calcium, magnesium and iron, if the insoluble substance was a mineral silicate.

Fusion with Potassium Hydrogen Sulphate

Any residue after the treatment above should contain only zirconium and titanium oxides. It can be fused with potassium hydrogen sulphate, the following reactions taking place:

$$TiO_2(s) + SO_3 \longrightarrow TiO^{2+} + SO_4^{2-}$$

$$ZrO_2(s) + SO_3 \longrightarrow ZrO^{2+} + SO_4^{2-}$$

Zirconium is detected by the precipitation of zirconium phosphate. Titanium, which also forms an insoluble phosphate, is detected by addition of hydrogen peroxide, which forms a yellow peroxotitanate. Formation of this complex prevents titanium phosphate precipitation.

REFERENCES

[1] Welcher, F. J. in Meites, L., ed. *Handbook of Analytical Chemistry*, McGraw-Hill, New York, 1963.

[2] Chalmers, R. A., *Aspects of Analytical Chemistry*, Oliver and Boyd, Edinburgh, 1968.

[3] Stephen, W. I., *Educ. in Chem.*, 1969, **6**, 221.

[4] Burns, D. T., Townshend, A. and Carter, A. H., *Reactions of the Elements and their Compounds*, Horwood, Chichester, in preparation.

[5] Burns, D. T., *Mikrochim. Acta*, **1967**, 147.

[6] Burns, D. T. and Drake, G. H., *Mikrochim. Acta*, **1967**, 389.

[7] Jones, W. F., *Mikrochim. Acta*, **1967**, 1004; **1959**, 635.

[8] Hodgman, C. D., ed., *Handbook of Chemistry and Physics*, Chemical Rubber Publishing Co., Cleveland, U.S.A.

[9] Firsching, F. H., *Talanta*, 1959, **2**, 326.

[10] Sinha, B. C. and Roy, S. K., *Analyst*, 1973, **98**, 289.

[11] Falkner, P. R. and Burns, D. T., *Mikrochim. Acta*, **1967**, 690.

[12] Belcher, R. and Farr, J. P. G., *Talanta*, 1959, **2**, 95.

[13] Ringbom, A., *Solubilities of Sulphides*, Report to Analytical Section of IUPAC, July, 1953.

[14] See Clifford, A. F., *Inorganic Chemistry of Qualitative Analysis*, Prentice-Hall, London, 1961.

[15] Dowson, W. M., *Mikrochim. Acta*, **1959**, 841.

[16] See, for example, Vogel, A. I., *A Text-book of Qualitative Chemical Analysis*, 3rd Ed., p. 27, Longmans, London, 1945.

[17] Palmer, A. H., *School Sci. Rev.*, **XLIII**, 735 (1962).

[18] Dowson, W. M., *Mikrochim. Acta*, **1969**, 202.

[19] Treadwell, W. D. and Schaufelberger, F., *Helv. Chim. Acta*, 1946, **29**, 1936.

[20] Kolthoff, I. M. and Griffiths, F. S., *J. Phys. Chem.*, 1938, **42**, 531, 541 and references therein.

[21] Caldas, E., *Anal. Chim. Acta*, 1969, **45**, 532.

[22] Beardsley, D. A., Briscoe, G. B., Clark, E. R., Matthews, A. G. and Williams, M., *Mikrochim. Acta*, **1970**, 1287.

[23] James, C. F. and Woodward, P., *Analyst*, 1955, **80**, 825.

[24] Burns, D. T., *Mikrochim. Acta*, **1965**, 920.

[25] See, for example, Moeller, T., *Qualitative Analysis*, McGraw-Hill, New York, 1958.

[26] Swift, E. H., *A System of Chemical Analysis*, p. 284, Prentice-Hall, New York, 1938.

[27] Stephen, W. I. and Weston, A. M., *Mikrochim. Acta*, **1964**, 179.

[28] Jones, W. F., *Mikrochim. Acta*, **1961**, 214.

[29] Hayes, O. B. and Winterburn, J., *Mikrochim. Acta*, **1958**, 197.

[30] Caldas, A. and Gentil, V., *J. Chem. Educ.*, 1958, **35**, 545.

[31] Dobbins, J. T. and Ljung, H. A., *J. Chem. Educ.*, 1935, **12**, 586; 1936, **13**, 75.

[32] Duschak, A. D. and Sneed, M. C., *J. Chem. Educ.*, 1931, **8**, 1177, 1386.

[33] Flosdorf, E. W. and Henry, C. M., *Ind. Eng. Chem., Anal. Ed.*, 1932, **4**, 434; *J. Chem. Educ.*, 1936, **13**, 274.

[34] Noyes, A. A., *J. Am. Chem. Soc.*, 1912, **34**, 609.
[35] Taimni, I. K. and Lal, M., *Anal. Chim. Acta*, 1957, **17**, 367.
[36] Weber, H. C. P. and Winkelmann, H. A., *J. Am. Chem. Soc.*, 1916, **38**, 2000.
[37] Welcher, F. J. and Briscoe, H. T., *Chem. News*, 1932, **145**, 161.
[38] Swift, E. H. and Schaefer, W. P., *J. Chem. Educ.*, 1961, **38**, 607.
[39] Belcher, R. and Weisz, H., *Mikrochim. Acta*, **1956**, 1847.
[40] Belcher, R. and Weisz, H., *Mikrochim. Acta*, **1958**, 571.
[41] Carter, A. H., *Mikrochim. Acta*, **1969**, 1097.
[42] Osborne, V. J. and Freke, A. M., *Mikrochim. Acta*, **1964**, 790.
[43] Belcher, R., *Analyst*, 1974, **99**, 802.
[44] See, for example, Newth, G. S., *A Manual of Chemical Analysis*, 6th impression, p. 184, Longmans, London, 1918.
[45] See, for example, ref. 16, pp. 403–409.
[46] Jones, W. F., *Mikrochim. Acta*, **1967**, 1019.
[47] Doležal, J., Povondra, P. and Šulcek, Z., *Decomposition Techniques in Inorganic Analysis*, Iliffe, London, 1968. '
[48] Bock, R., *A Handbook of Decomposition Methods in Analytical Chemistry*, trans. I. L. Marr, International Textbook Co., Glasgow, 1979.
[49] Stephen, W. I. and Thomas, D., unpublished work.
[50] Harpham, W. E., *Metallurgia*, 1955, **52**, 45.

Chapter 4

Techniques and Apparatus for Semi-Micro Qualitative Analysis

The practice of qualitative inorganic analysis enables the student to acquire a considerable knowledge of the properties and behaviour of inorganic ions and individual substances. Also, as with other practical sciences, to achieve success, he must develop manipulative skills. Facility with semi-micro techniques, in which solids are measured in milligrams and liquids by drops, develops manipulative skills of a high order and has great advantages over obsolescent larger-scale methods.

The following sections, which deal with the apparatus and techniques used in semi-micro analysis, are arranged from the point of view of the student beginning the examination of a typical powder and proceeding through to the final stage, where the results of the analysis are reported.[†] Detailed procedures follow in Chapter 5.

The Sample for Cation Analysis

The weight of the sample for the cation analysis should be about 10 mg, though satisfactory analyses can be performed with samples as small as 5 mg or as large as 20 mg. Visual estimation of a 10-mg quantity is extremely difficult, and it is recommended that a sample of about 30 mg be dissolved in a suitable solvent to give about 3 ml of solution. One ml of the prepared solution can then be taken for the cation analysis.

It has been found that the end of a nickel semi-micro spatula, moderately heaped with the powdered material, will generally give a sample of 15–30 mg (Fig. 4.1). Salts differing in density as widely as lead nitrate and calcium carbonate fall within this range [1].

[†] A movie film (*ca.* 40 min.) describing the use of the apparatus and techniques has been produced by MAQA, and is available for hire from Dr. E. R. Clark, Chemistry Department, University of Aston, Birmingham B4 7ET, U.K.

Fig. 4.1 – Spatula with sample.

Use of 5-ml Beaker

The solution for cation analysis is prepared in a 5-ml beaker. The beaker should be not more than two-thirds filled with liquid. If heating is necessary, this is done on the heating block (p. 151). Mixing of the solution is done most effectively by squeezing and releasing the teat of the teat pipette (p. 150) with the tip of the pipette immersed in the solution. This method of mixing is much more efficient than conventional stirring with a glass rod and there is less danger of upsetting the beaker. The 3 ml of solution for analysis are prepared in the beaker and about 1.0 ml is transferred with the teat pipette to a centrifuge tube; the rest of the solution is stored in another tube in case it is necessary to repeat the analysis.

If the sample is not completely soluble in the usual solvents (p. 174), the *residue* is transferred (p. 156) to a suitable vessel and fused with the appropriate reagent. Two vessels, a *porcelain crucible* and a *nickel spoon,* are recommended as able to withstand the high temperatures and pronounced chemical reactivity common to these fusion procedures. Acid fluxes attack nickel, alkaline fluxes attack porcelain; hence attention must be given to using the correct vessel with the appropriate reagent.

Use of Porcelain Crucible

The 5-ml porcelain crucible is used for fusions with acidic reagents, for example, potassium hydrogen sulphate. The sample is placed in the bottom of the clean dry crucible with about ten times its weight of the acidic flux. The contents of the crucible are then fused by placing the crucible on a pipe-clay triangle and heating it directly with the full flame of a microburner (see p. 151). The mixture is kept molten for $\frac{1}{2}$-1 minute, until the sample has disappeared into the melt. The burner is then withdrawn and the crucible is carefully tilted to allow the melt to solidify over as much of the inner surface of the crucible as possible, thus simplifying the subsequent dissolution of the fused mass and reducing the risk of cracking the crucible. When quite cool, the crucible is removed from its support, 2-3 ml of water are added and complete dissolution is effected by gentle heating of the crucible on the heating block.

Use of Nickel Spoon

The nickel spoon (Fig. 4.2) is used for fusions with *alkaline* reagents, for example, sodium carbonate or fusion mixture. The sample is placed in the spoon with about ten times its weight of the alkaline flux. The spoon, held by the tongs, is placed above the flame of a microburner and heated until the mixture melts and attack on the sample is complete. After it has cooled sufficiently, the spoon is placed in the 5-ml beaker, sufficient water or other solvent is added, and the fused mass is leached from the spoon by gently boiling the contents of the beaker on the heating block.

Fig. 4.2 – Nickel spoon.

Use of Tongs

The tongs should be made of nickel and so designed that they can be used to lift the beaker or centrifuge tube from the hot block, and also the crucible from the block or pipe-clay triangle. They should also be capable of holding the nickel spoon during fusions. If they have bent tips, they should always be placed on the bench with the tips uppermost; the wrist can be turned so that the tips point downwards for use.

Glass Stirring Rods

The glass stirring rods should be 10 cm long, 1-2 mm in diameter, with one end rounded and the other end cut flat but *not* fire-polished. The flat end may be used to dispense drops from centrifuge tubes onto test-papers.

Centrifuge Tubes

The most important operation in inorganic qualitative analysis is the separation of a precipitated solid from a liquid. In semi-micro work this is done by means of the centrifugal force generated in a centrifuge, which forces the solid to the bottom of the containing tube – the centrifuge tube – and leaves a clear liquid above. The clear liquid is separated from the solid with a teat pipette.

Suitable centrifuge tubes are 7.5 cm long and 1 cm outside diameter (about 3.5 ml capacity). It is advisable to have fairly thick-walled tubes, otherwise breakage may easily occur. The technique of centrifugal separation replaces the filter funnel, filter paper and test-tube of the older practices. Apart from operations such as fusions, evaporations and gas tests, the analysis is largely carried out in centrifuge tubes.

One or two centrifuge tubes should be calibrated at 0.5, 1.0, 1.5, 2.0 and 3.0 ml. It is sufficiently accurate to add these volumes of water from a burette. The calibration marks should be made with a file.

In general, a centrifuge tube should never be more than half filled with solution. If the volume becomes greater than this, the solution should be transferred to the beaker and reduced in bulk by evaporation (p. 152). A solution in the tube is best mixed either by use of a stirring rod or by squeezing and releasing the teat of a teat pipette (below) while the tip of the latter is immersed in the solution. Occasionally it may be necessary to close the mouth of the tube with a bung and shake vigorously to obtain mixing.

It is most important to keep all centrifuge tubes scrupulously clean (p. 161).

Teat Pipette

A solution is transferred from one vessel to another by means of a teat pipette. The recommended type of teat pipette is shown in Fig. 4.3. The delivery tube below the bulb is of a predetermined length and diameter to ensure a uniformity of drop size for any given liquid. The bulb minimizes suck-back of liquid into the rubber teat and is so situated that it fits into the palm when the teat is held between the thumb and forefinger of the right hand.

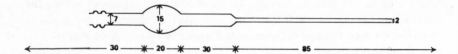

Fig. 4.3 – Teat pipette (dimensions in mm).

As exact volumes often have to be dispensed from the teat pipette it must be calibrated. Drops of water are added from the pipette to a calibrated centrifuge tube (above) until the 1.0-ml mark is reached. The number of drops per ml is thus obtained and the volume of a single drop is calculated. It is essential that the calibrated pipette dispenses *12–20 drops per ml, and that the volume of a single drop is not greater than 0.09 ml.*

When the teat pipette is used to transfer a solution, the teat should be squeezed before the tip of the pipette is placed in the solution. This avoids inadvertently adding small amounts of residual liquid which may be present, even in a 'cleaned' pipette, to the solution and causing dilution or contamination. Moreover, a clear solution can thus be removed from above a centrifuged precipitate, without disturbing the precipitate.

It is essential that the pipette be thoroughly cleaned after each operation in which it has been used. Concentrated hydrochloric acid may be used for dissolving most sulphides and basic oxides. A small amount of bromine in the hydrochloric acid frequently expedites dissolution, but if the pipette is *washed immediately after each operation* little trouble will be experienced. To wash the pipette, the teat is detached, and water is allowed to run through the inside and over the outside of the pipette; finally it is rinsed with distilled water.

A useful plan is to place the used pipette, with the teat detached, in a tall-form beaker through which running water flows. If, say, three pipettes are used in this way, the cleaning operation is considerably eased.

Use of Heating Block

A *heating block* (Fig. 4.4) is the most suitable means of heating centrifuge tubes and 5-ml beakers. The most common type of heating block consists of a cylinder of aluminium alloy mounted on three legs. Four holes are drilled in the block. The centre one accommodates a small stout-walled thermometer to read temperatures of up to 200°. Two of the other holes accommodate beakers and the remaining one may be used for a centrifuge tube. One of the larger beaker holes can also be used to hold the porcelain crucible. The base of each hole is lined with a pad of ceramic wool.

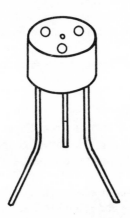

Fig. 4.4 – Heating block.

A *microburner* is used to heat the block; specially designed combinations of burner and heating block are also available. It is recommended that a burner with a controlled air inlet be used rather than one with a fixed air inlet. If it is not possible to control the air, the flame is likely to become extinguished when the gas pressure fluctuates, or when an attempt is made to control the temperature of the heating block by turning the flame down.

The burner and heating block should be placed on a heat-resistant or asbestos board. This prevents damage to the bench by heat, or by material that may be spattered from vessels in the block.

When the small burner is placed underneath the heating block, the latter usually reaches its maximum temperature in 15–20 minutes. The temperature should be checked on the thermometer and controlled so that the block does not become too hot. A suitable temperature is 120–130°, but not higher. It will soon be appreciated what size of flame is suitable to achieve this range of temperatures. Draughts, which disturb the small flame, are overcome by shielding the burner with aluminium foil wrapped round the vertical part of the tripod legs, a small gap being left at the top for the escape of burned gases.

When a centrifuge tube containing liquid is heated to boiling point on the block, it must be lifted frequently and rotated rapidly with the fingers as soon as it reaches the stage where bubbles appear in the liquid, to avoid ejection of solution by too rapid onset of boiling. *Prolonged boiling should be carried out only in the beaker and not in the centrifuge tube.* In most instances, the boiling of materials in the beaker by using the heating block is a simple technique. Smooth boiling can often be achieved by gently blowing air through the solution with the teat pipette, just before boiling begins. This operation will also assist the expulsion of dissolved gases. Gentle stirring of the solution with a glass rod will also effectively control bumping.

Operations at lower temperatures, such as evaporation of the solution to dryness in a beaker or crucible, can be carried out by placing the vessel on the surface of the block. Evaporation can be speeded up by directing a *gentle* stream of air from a capillary onto the surface of the liquid (just enough to dimple it).

Alternative Methods of Heating

It is sometimes expedient to heat a centrifuge tube directly by means of a microburner. The tube is held in the pair of tongs, which has been specially designed to hold it firmly. To prevent bumping, the tube is *held just above the tip of the microburner flame and shaken carefully.* The tube should not be placed in the flame nor should it ever remain still. It is advantageous to place a stirring rod in the tube. The rod should remain loose but not be moved more than can be helped.

Centrifuge tubes can also be heated by placing them in a 100-ml beaker containing boiling water. Advantages are that bumping seldom occurs and the tubes may be left unattended.

High-temperature fusions are carried out in a crucible which is heated directly over the burner. The crucible is held in the tongs or supported on a pipe-clay triangle placed across the top of a tripod.

These methods of heating vessels may be used in addition to the heating block.

Wooden Block for Apparatus

It is convenient to have a block for storage of some of the apparatus. A suitable block (Fig. 4.5) measures 9 × 20 cm with a depth of 3 cm. There are two rows of seven holes each along the back of the block to accommodate the centrifuge tubes. The holes are so arranged that the tubes in the back row stand higher than the tubes in the front row. At each end, between the line of the two rows, there is a further hole of similar size. One of these is used to store the sample tube or the tube containing the platinum wire used in flame tests. The other is used to store the centrifuge tube which is used as a counterpoise (p. 155). The front row is used to store clean tubes and the back row is used to set aside centrifuged precipitates belonging to the various groups, before these are analysed.

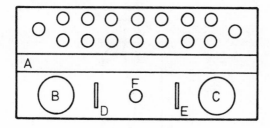

Fig. 4.5 – Plan view of apparatus block.

Immediately in front of the second row lies a trough-shaped depression, 'A'. This is used to store stirring rods, the semi-micro spatula, possibly the platinum wire, and the teat pipette. In front of this depression there are two large holes, 'B' and 'C', lying near each end of the block, to hold beakers. In advanced work where fusions are sometimes necessary, these holes hold a beaker and a porcelain crucible (p. 148). Slits 'D' and 'E' serve to hold 2.5-cm watch glasses in an upright (edgewise) position. The depression 'F' holds a nickel spoon (p. 149), again necessary only in advanced work. Obviously, there can be many variations in the design of the wooden block. The one described satisfies most of the requirements for both elementary and advanced work.

Addition of Reagents

All the reagent solutions used in this scale of analysis are conveniently stored in dropping bottles of the ground-glass teat pipette type (Fig. 4.6). Ordinary ground-glass stoppered bottles are not generally recommended, because of the risk of contaminating the reagents with a badly cleaned teat pipette. The storage of strong acids and bromine water presents certain difficulties because these reagents attack the rubber teats of the dropping pipettes. Teats made of vulcanized polythene should be used with these reagents (concentrated nitric acid attacks this material to a slight extent and it is recommended that the teat on the nitric acid bottle be removed, washed in water and replaced, when this reagent has been used). Strong alkali tends to cement the pipette to the neck of the bottle; this can be avoided by smearing the ground joint with silicone grease. Alternatively, such reagents can be dispensed from small polythene dropping bottles. The dropping pipettes should deliver about 20 drops per ml. When a reagent is added to a centrifuge tube from a dropping pipette, care must be taken to avoid contact of the end of the pipette with the inside wall of the tube. *Failure to do this can seriously contaminate the reagents.*

Fig. 4.6 – Dropping bottle.

Reagents which are much more or much less dense than the solution being treated must be added carefully. They tend to form a layer at the bottom or top of the tube, respectively, and it is always advisable, therefore, to mix the solution thoroughly after the addition of each reagent, by gently blowing air through the solution by using the teat pipette with its tip situated at the bottom of the tube. In this way, the whole contents of the tube are thoroughly mixed.

Throughout the analytical scheme there are many occasions when it is necessary to make a solution either acidic or ammoniacal. The use of small squares of litmus paper in these operations is useful. These are obtained by cutting the normal strips of litmus (as supplied in book form) lengthwise down the middle and then tranversely to give the appropriately sized squares. The addition of the correct amount of ammonia to a solution is troublesome; there is usually sufficient ammonia in the vapour phase above the solution to give an appreciable smell or reaction with moist litmus although the actual solution may still be quite acidic. The reverse procedure, that of acidifying an ammoniacal solution, is subject to similar difficulties. Putting the litmus paper into the solution overcomes these problems.

Solid reagents should always be added from the tip of the nickel spatula.

Hydrogen sulphide is added as a solution in acetone [2]. At room temperature, 1 litre of acetone dissolves about 20 g of hydrogen sulphide. This solution is stable for at least three months when it is prepared as described (p. 236). As the saturated solution in acetone has a comparatively high concentration of hydrogen sulphide, only a few drops of the reagent need be added to effect complete precipitation in the Copper–Tin and Zinc Groups. Acetone dissolves in aqueous solutions, and is eventually boiled out during further operations. It is important that the hydrogen sulphide in acetone reagent should not be contaminated by acid; hydrogen ions promote rapid formation of thioacetone with consequent loss of the sulphide-precipitating properties of the reagent. It is therefore desirable to test the effectiveness of the reagent periodically with a copper solution.

Hydrogen sulphide is highly toxic and therefore must always be used in a fume cupboard.

Use of the Centrifuge

Precipitates are separated from solutions by centrifuging at 2000–3000 rpm. Before this operation, a similar tube is filled with water to the same level as that of the solution in the tube containing the precipitate. This is the *counterpoise tube* which is placed in the bucket at the end of one arm of the centrifuge, the sample being placed at the other end. It is a distinct advantage to reserve one labelled tube as the counterpoise for all operations. It will then not be confused with the tube containing the test solution, as may happen if the test solution appears clear after centrifuging, to ascertain whether or not a slight precipitate has been formed. When properly balanced a centrifuge will run quietly, with very little vibration. If it is not properly counterpoised, a vibration will be set up. If this occurs, it should be switched off immediately, and the level of water in the counterpoise tube adjusted. Severe damage may result to the centrifuge if it is used when not properly counterpoised. It is usually sufficient to spin the centrifuge for only about ten seconds by holding the switch down for that period of time. Very finely divided precipitates and gelatinous precipitates may take a little longer to settle. The process can often be expedited by adding a few drops of ethanol, which is also particularly effective in loosening any precipitate which adheres to the wall of the tube.

Separation and Washing of a Precipitate

A precipitate is separated by placing the tip of the teat pipette, with the teat depressed, near the solid material at the bottom of the tube. Slow release of the teat causes the supernatant liquid to rise into the body of the pipette without disturbing the precipitate. In this operation, the tube containing the precipitate and liquid to be separated is held between the thumb and first finger and the tube into which the liquid is to be transferred between the first and second fingers of the same hand. The tubes are held at 45° above the horizontal to ease the removal and transfer of the liquid. The tube containing the transferred solution is then placed in position in the wooden block.

The precipitate is now ready to be washed. This is done by adding about 1 ml of wash liquid to the tube and stirring up the precipitate either with a thin glass rod or with air from the teat pipette. The suspended precipitate is centrifuged once more, the level of the liquid in the counterpoise tube being adjusted to match that of the wash liquid in the other tube, if necessary. The wash liquid is removed as described above, and is either rejected or retained, depending on whether it is required for further tests. One washing in this way is normally sufficient, but it is recommended to use two.

The final operation is the draining of the precipitate. The transfer of liquid as described above generally leaves a small amount of liquid in contact with the precipitate, and this cannot successfully be removed by means of a teat pipette without disturbing the precipitate. This is overcome by inclining the tube about 30° below the horizontal so that the residual liquid drains towards the mouth of the tube and gathers round the lip; this liquid is easily removed by means of the teat pipette, or, if it is not required for further examination, by means of a small piece of filter paper.

Transfer of a Precipitate

It is occasionally necessary to transfer part of a precipitate to another tube in order to carry out a particular test. This is done by suspending the precipitate in a suitable liquid, generally water, and withdrawing a sufficient volume of the suspension with the teat pipette. This portion of the suspension is transferred to another tube, and if the levels of liquid in the two tubes are made equal, both can be centrifuged at the same time without the need for the usual counterpoise tube. The precipitates are then separated and drained as described below. When a sample for cation analysis is not completely soluble in the usual solvents and an insoluble residue is obtained which must be fused (p. 139) the residue may be transferred to the fusion vessel as follows. The residue is first washed and drained well. Sufficient of the dry, powdered fusion reagent is added to the tube and is well mixed with the residue. This mixture is transferred to the fusion vessel (crucible or nickel spoon) by means of the micro-spatula. A more general procedure is to wash the residue in the tube with ethanol, drain it well and evaporate the residual solvent by placing the tube in the heating block for a few minutes. The dry residue is easily tapped out into the fusion vessel and subjected to the appropriate treatment.

Testing for Gases

One of the most troublesome operations in semi-micro analysis is the identification of a gas which may be evolved. Several pieces of apparatus are available commercially but, in general, they are not very satisfactory.

Some gases (for example, ammonia, hydrogen sulphide, sulphur dioxide) can be detected by using reagent papers, but contamination of the reagent paper by spray from the reactions must be avoided, as, for example, when

Silicate ions can also be identified by this method. The test sample is first mixed with sodium fluoride and the silicon tetrafluoride is liberated by treatment with concentrated sulphuric acid.

Hexafluorosilicate can be distinguished from fluoride or from silicate, since silicon tetrafluoride is liberated on heating the test material alone with concentrated sulphuric acid.

Fig. 4.9 – Lead plate assembly.

Flame Tests

Flame tests are particularly valuable for revealing the presence of alkali and alkaline earth metals. The platinum wire must be cleaned by heating in a nonluminous flame after dipping into concentrated hydrochloric acid, the operation being repeated until no persistent colour is imparted to the flame. Usually a faint golden colour, which persists for a moment in the flame, is obtained owing to the presence of trace amounts of sodium. The solid to be examined is placed on a small watch glass, the wire is again moistened with concentrated hydrochloric acid and its end only is touched against some of the solid. The wire is then heated in the flame and any characteristic colours noted. It is sometimes possible to detect one metal in the presence of another by using a screen of coloured glass to filter out unwanted regions of the spectrum. Thus the lilac potassium flame can be seen through a blue (cobalt) glass even when the flame is deep yellow because of the presence of sodium.

The wire is moistened with hydrochloric acid so that the solid may adhere, and also to ensure that some of the metal ion is present as its chloride. An adequate concentration of salt in the flame is then more likely to be obtained because chlorides are usually more volatile than other salts.

The platinum wire (32 s.w.g.), about 3 cm long, is usually sealed into a 3-mm diameter glass rod for ease of handling. The rod and wire should be stored in an upright position in a centrifuge tube with the wire dipping into hydrochloric acid so that the wire stays clean.

The best method of cleaning a platinum wire is to dip the moistened wire in finely ground potassium hydrogen sulphate, fuse the adhering material in the burner flame, then wash the wire in water.

Where platinum is not available, a nichrome wire or an ordinary pencil with the wood shaved well back from the graphite may be used. The method of use is similar and the results are surprisingly satisfactory.

Evaporation to Fumes

When it is necessary to evaporate a liquid to small volume or to fumes, the operation should be carried out in the beaker (or crucible), not in a centrifuge tube. When a tube is used, there may be losses because of spurting; furthermore the narrow walls of the tube act as a condenser, and complete evaporation is difficult to achieve.

Sometimes copious fumes are evolved in these evaporations. A very useful device to overcome this unpleasant feature, if a fume cupboard is not available, is to use an inverted filter funnel (10–12 cm diameter) connected to the water-pump by rubber tubing. If this is situated above the heating block and gentle suction is applied, acid or ammonia fumes are easily led away (Fig. 4.10).

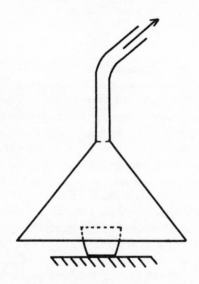

Fig. 4.10 – Arrangement for removal of fumes.

Clean Working

Clean and tidy working is essential for correct results. Figure 4.11 shows a suggested layout for a laboratory bench.

Behind the apparatus block are two 250- or 400-ml beakers containing distilled or demineralized water. The teat pipette is rinsed under the tap, washed with water from one of the beakers and stored in the other beaker. It is important to change the rinse water frequently.

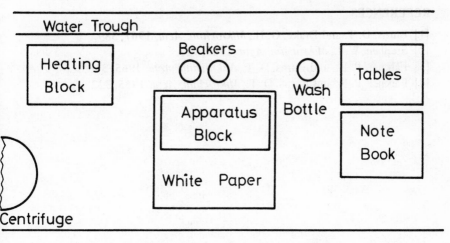

Fig. 4.11 – Suggested lay-out for laboratory bench.

Cleaning of Glassware

All glass apparatus must be kept scrupulously clean. Inorganic precipitates adhering to the walls of the tubes can be removed with a centrifuge tube brush or a suitable solvent. Grease is perhaps the most troublesome variety of dirt and is generally best removed by soaking the vessel for some time in an efficient detergent solution. Sometimes it is simpler and speedier to use the centrifuge tube brush and a mildly alkaline abrasive powder to clean the walls of the vessel.

The most frequently used piece of apparatus is the teat pipette. Directions have already been given for cleaning the pipette (p. 150) but the importance of having a clean pipette cannot be over-emphasised. No matter what method is used to clean the pipette, *this must be done after each operation in which it is used.*

Writing of Results

Each teacher has a preferred method for writing up the results of laboratory work. Whichever method is used, it is strongly recommended that the results are written *immediately* after making the experimental observation. Otherwise, it is not possible, except with outstanding students, to obtain an accurate report. A student may very often form strongly defined opinions about the nature of his sample and may ignore or adapt earlier observations to fit a particular hypothesis. The golden rule of all experimental work is 'Write it down at once'. Extensive detail is unnecessary and should be avoided.

REFERENCES

[1] Burns, D. T. and Drake, G. H., *Mikrochim. Acta,* 1967, 389.
[2] Stephen, W. I., *Mikrochim. Acta,* 1960, 927.
[3] Falkner, P. R., and Burns, D. T., *Mikrochim. Acta,* 1965, 318.
[4] Falkner, P. R. and Burns, D. T., *Mikrochim. Acta,* 1965, 322.

Chapter 5

Systematic Semi-Micro Qualitative Inorganic Analysis - Experimental Procedures

This Chapter describes the procedures recommended by MAQA [1] for the systematic qualitative analysis of an unknown sample. The chemicals and solutions required are given in the Appendix (p. 233).

5.1 OUTLINE OF PROCEDURE FOR THE EXAMINATION OF AN UNKNOWN MIXTURE

The systematic anion separation scheme is linked with the preliminary anion examination. It is usual for the cation analysis to be completed before the systematic anion analysis.

A few preliminary tests, which often give useful indications of the nature of the sample, should be carried out first. Because certain anions interfere in the general cation scheme, it is essential that the presence or absence of these be established before starting the cation analysis. The use of smell is particularly important in these tests, and proper use of smell can save much time. However, some caution is necessary, for most of the evolved gases are toxic, and some are *very toxic* (hydrogen cyanide, hydrogen sulphide, carbon monoxide). Repeated smelling, which will in any case dull the sense of smell, should be avoided. The following general procedure is recommended.

1. Note the colour and uniformity of the sample.
2. Begin the extraction with sodium carbonate solution (p. 171). It takes some time to prepare and some will be required for the preliminary anion examination. During the boiling, test for ammonia and reserve the prepared solution for the potassium and the anion tests.
3. Whilst the sodium carbonate extract is being prepared, carry out the preliminary dry tests (p. 164).
4. Carry out the preliminary anion tests on the sodium carbonate extract, and establish the presence or absence of anions which interfere with the cation scheme (p. 170).
5. Prepare a solution for the cation analysis (p. 174).
6. Carry out the cation analysis (p. 176).
7. Carry out the anion analysis (p. 191).

5.2 ANALYSIS OF MIXTURES CONTAINING COMMON ANIONS AND CATIONS

5.2.1 Preliminary Dry Tests

The preliminary dry tests can give very useful information about the composition of a substance. Nevertheless the time devoted to them should be relatively short, for in general they give indications only; proper identification is carried out later.

The charcoal block, borax bead and microcosmic salt bead reactions are better omitted for simple compounds, but they are particularly valuable in the identification of insoluble substances (p. 213).

TEST	RESULT	INFERENCE
COLOUR OF MIXTURE	Colourless	Copper, iron, chromium, cobalt, nickel, manganese, permanganate, chromate, hexacyanoferrate(II) and hexacyanoferrate(III) ions are probably absent.
	Yellow	May be hydrated iron(III) salt, chromate or dichromate.
	Pale pink	Hydrated manganese(II) salt.
	Rose red	Hydrated cobalt(II) salt.
	Pale green	Hydrated iron(II) salt.
	Dark green or purple	Hydrated chromium(III) salt.
	Green	Hydrated nickel(II) salt.
	Green or blue	Hydrated copper(II) salt.
	Deep blue	Anhydrous cobalt(II) salt.
EFFECT OF HEAT Heat about 0.1 g of the mixture in a dry ignition tube held almost horizontally in the burner flame.	Substance BLACKENS	Organic substance, e.g. tartrate or salt of metal of which the oxide is black: copper, iron, cobalt, nickel.
	A SUBLIMATE FORMS: White	Ammonium salt, arsenic.
	White, after solid melts	Mercury(II) chloride, aluminium chloride.
	Yellow—hot, white—cold	Mercury(I), antimony.
	Grey or black giving globules when rubbed	Mercury
	Purple–black; violet vapour	Iodine from certain iodides.
	GAS EVOLVED: Odourless, colourless, inflammable	Carbon monoxide from oxalate.
	Odourless, colourless	Carbon dioxide from organic acid, carbonate, hydrogen carbonate, hydrated complex cyanide. Oxygen from certain oxides, peroxides, alkali metal nitrates, bromates, iodates, chlorates.
	Foul smelling, colourless, inflammable	Phosphine from phosphite, hypophosphite; hydrogen sulphide from certain hydrated sulphides.

TEST	RESULT	INFERENCE
	Colour	
	Greenish-yellow	Chlorine from chloride in presence of an oxidizing agent.
	Reddish-brown	Nitrous fumes from certain nitrates, or bromine from bromate or bromide in the presence of an oxidizing agent.
	Violet	Iodine from certain iodides and iodates.
	Odour	
	Sulphurous	Certain sulphites and sulphates.
	Ammoniacal	Ammonium salt, hydrated complex cyanide.
	Action on damp litmus paper:	
	Blue to red, acid vapours	Certain sulphates, nitrates, hexafluorosilicates.
	Red to blue, alkaline vapours	Ammonium salts, hydrated complex cyanides.
	Bleached	Chlorine or bromine from: (a) certain chlorides or bromides. (b) chloride or bromide in the presence of an oxidizing agent.

TEST	RESULT	INFERENCE
FLAME TEST Heat the wire low down in the edge of the non-luminous flame, gradually raising it to the hottest point above the inner cone.	Persistent yellow; completely masked by cobalt glass	Sodium
	Lilac; shows crimson through cobalt glass	Potassium
	Apple green	Barium
	Blue–green	Copper or borate
	Brick-red; shows light green through cobalt glass	Calcium
	Crimson; almost unchanged through cobalt glass	Strontium
	Carmine red	Lithium

Other elements may give flame colours. For example, tin gives a greyish-blue colour, lead, arsenic and antimony give bluish-violet colours, and zinc sometimes gives a pale green colour. These colours are, however, of little diagnostic value.

CHARCOAL BLOCK
REACTION [2]

TEST	RESULT	INFERENCE
(Recommended only for *insoluble substances*). Make a small depression in the block (not a deep, narrow hole), and place in it 20 mg of the sample. Add twice the bulk of anhydrous Na_2CO_3 and mix. Heat with a blowpipe or with a glass tube having a jet hole (26 S.W.G.) connected directly to the gas supply. If decrepitation occurs exercise great care.	INCRUSTATION formed:	
	Yellow – hot, white – cold	Zinc
	White; garlic odour	Arsenic
	Slightly yellow – hot, white – cold	Tin
	White	Antimony
	Yellow	Bismuth
	Chocolate brown, some distance from flame	Cadmium
	Dark red-brown – hot, yellow – cold	Lead

TEST	RESULT	INFERENCE
CHARCOAL BLOCK (contd.)	METALLIC BEAD:	
	Hard white	Silver
	Soft white	Tin
	Soft white; marks paper	Lead
	Greyish and brittle	Bismuth, antimony
	Red scales or flakes	Copper
If the presence of zinc, aluminium, or magnesium is suspected, heat the substance alone on the block for 10-15 sec; allow it to cool, moisten it with 1 drop of $Co(NO_3)_2$ solution and heat	Green mass	Zinc
	Blue mass	Aluminium, fusible phosphates, arsenates, borates, silicates.
	Pink mass	Magnesium

Detection of Sulphur in an Insoluble Sulphate

Insoluble substances such as barium, strontium and lead sulphates are reduced to sulphides when heated on the charcoal block. The sulphide formed is detected as follows.

A. Transfer the residue to a piece of bright silver foil and moisten with a drop of water. A brown or black stain is produced. Alternatively a cupronickel coin may be used, but the test is then slightly less sensitive.

B. Transfer the residue to the crucible, add 2-3 drops of dil. hydrochloric acid, and test the evolved gas for H_2S with lead acetate paper.

BORAX BEAD TEST

Form the end of the platinum wire into a small loop and heat it in the burner flame until red hot. Quickly dip into powdered borax and heat the salt in the hottest part of the flame to drive off water and form a colourless transparent bead. Cool. Moisten the bead and bring into contact with the powdered mixture so that only a minute amount adheres. Heat in the normal oxidizing flame (O) and then in the reducing flame (R) obtained by closing the air vent until there is a small yellow tip to the flame.	O. Brown ⎱ R. Grey ⎰	Nickel
	O. Brown – hot, Yellow – cold ⎫ R. Bottle-green – hot Yellow-green – cold ⎭	Iron
	O. Blue ⎱ R. Blue ⎰	Cobalt
	O. Green – hot, blue – cold ⎫ R. Colourless – hot, red – cold ⎭	Copper
	O. Emerald green ⎱ R. Emerald green ⎰	Chromium
	O. Amethyst (pale purple) ⎱ R. Colourless ⎰	Manganese
Chromium and manganese may be further identified as follows. Form a sodium carbonate bead, using anhydrous Na_2CO_3, and then introduce a small amount of KNO_3. Cool. Moisten, and bring into contact with the powdered mixture. Heat strongly.	Yellow bead	Chromium
	Dark green bead (if very dark add more sodium carbonate)	Manganese

MICROCOSMIC SALT BEAD Produce a bead with microcosmic salt [$Na(NH_4)HPO_4.4H_2O$] in a similar manner to that described for the borax bead. This test is recommended only for the detection of silicate.	Colours of beads are similar to borax beads. Many silicates give a semi-translucent mass of silica suspended in the bead, and this silica 'skeleton' can be seen during and after fusion.	Silicate

5.2.2 Preliminary Non-Systematic Examination for Anions

Before beginning the cation separation scheme, it is essential to test for certain anions which, if present, might interfere in the cation separations. Although this may be accomplished, to a large extent, by using the systematic anion scheme (p. 191), it may be more convenient at this stage to apply the selective tests described immediately below. The interfering anions (p. 127) are borate [3], fluoride, hexafluorosilicate, organic anions, phosphate, silicate and thiosulphate.

Tests for all these ions except phosphate are described below. In addition, it is convenient at this stage to test for potassium and ammonium ions, specified organic anions and cyanide ions. Finally, it is necessary to establish whether arsenite or arsenate ions are present, in order to apply the appropriate modification to the cation separation scheme.

First, apply the general tests below, which give further indications of the anions present in the sample. Then apply the more selective tests for particular anions, for oxidants and for potassium and ammonium ions, as given on pp. 170-173.

5.2.3 General Tests

Use the apparatus and techniques described on p. 156 for the detection of gases.

TEST	RESULT	INFERENCE
EFFECT OF WATER		
Shake about 50 mg of sample with 1 ml of water.	Purple solution	Permanganate
	Yellow solution	Chromate
	Orange solution	Dichromate
Observe whether any gas is evolved and smell the mouth of the tube. Test any gas evolved.	Gas spontaneously combustible	Phosphine from phosphides of alkali and alkaline earth metals.
	Smell of H_2S; lead acetate paper blackened	Hydrogen sulphide from a few sulphides; e.g. those of magnesium and aluminium.
	Smell of SO_2; iodine–starch solution decolourized	Dithionite
	Gas relights glowing splint	Oxygen from alkali metal peroxide.
Heat the solution and test for evolved gases	Smell of O_3; starch–iodide paper turned blue	Ozone from peroxodisulphate.
	Smell of NH_3; moist red litmus paper turned blue	Ammonia from magnesium nitride or alkaline earth metal nitride.
	Gas relights glowing splint	Oxygen from barium peroxide.

TEST	RESULT	INFERENCE
EFFECT OF WATER (*contd.*)		
Add small pieces of red and blue litmus paper to the solution remaining in the tube.	Alkaline reaction	Alkali metal hydroxide or a peroxide; salt of a strong base/weak acid, e.g. alkali metal carbonate, borate or phosphate.
	Acid reaction	Peroxodisulphate, organic acid, acid salt or readily hydrolysed salt of strong acid, e.g. iron(III) or aluminium salt.
EFFECT OF DIL.HCl Place about 50 mg of sample in a centrifuge tube and add 0.5 ml of 4M HCl. Warm, if no apparent reaction. Smell the mouth of the tube and test any evolved gas with blue litmus paper. Test gas for ignition.	Gas ignites with slight explosion	Hydrogen from certain metals, e.g. aluminium, iron, zinc.
	Immediate orange colour which soon fades; SO_2 evolved and S deposited	Dithionite
	Smell of H_2S; lead acetate paper blackened	Hydrogen sulphide from sulphide; sulphur deposited in tube indicates polysulphide.
	Smell of Cl_2; pale green gas evolved; litmus paper bleached; blue colour with starch-iodide paper	Hypochlorite or oxidizing agent.
	Smell of Br_2; red–brown gas evolved; litmus paper bleached; blue colour with starch-iodide paper	Hypobromite, or bromide + strong oxidizing agent.
	Red–brown acidic gas evolved; blue colour with starch–iodide paper	Nitrous fumes from nitrite
	Smell of vinegar; moist blue litmus paper turned red	Acetic acid from acetate; confirm by test on p. 170.
Test with iodine–starch solution. If sulphide is present add 2 drops of satd. $HgCl_2$ solution to the sample before treating with dil. acid.	Smell of SO_2 on warming; iodine–starch solution decolourized	Sulphur dioxide from sulphite or thiosulphate, sulphur deposited in tube indicates thiosulphate or dithionite; these reactions should be sufficient to detect thiosulphate.
Test for CO_2 (p. 158). (If sulphite is present, treat the solution with 2 or 3 drops of 20 vol. H_2O_2)	Lime-water turned milky	Carbon dioxide from carbonate, hydrogen carbonate, or cyanate; confirm by tests on p. 197.

TEST	RESULT	INFERENCE

EFFECT OF CONC. H_2SO_4

If permanganate is suspected to be present, this test should be omitted

TEST	RESULT	INFERENCE
Add about 0.5 ml of conc. H_2SO_4 to approx. 50 mg of the sample in a centrifuge tube. If any gases are evolved test with blue litmus paper and place the mouth of the tube at the edge of the bunsen flame to see if the vapours will ignite. Then warm the tube and, exercising care, gradually heat more strongly. Note the colour of any gas evolved, whether any charring occurs, and the appearance of the tube. Test any gas evolved with blue litmus paper and if a colourless gas is evolved test for CO_2 by the lime-water reaction.	Gases evolved by treatment of the sample with dil. HCl will be liberated more vigorously with conc. H_2SO_4. Other effects giving useful information are as follows.	
	1. Yellow gas in the cold, or on gentle warming (with crackling in the solution). This reaction is very explosive and if there is any evidence of a chlorate the treatment of the sample with conc. H_2SO_4 must be abandoned — stop warming and carefully pour into a large volume of cold water.	Chlorine dioxide from chlorate.
	2. Misty acid fumes giving a turbidity on a rod previously dipped into $AgNO_3$ solution	Hydrogen chloride from chloride, hydrogen bromide from bromide.
	3. Violet vapours	Iodine from iodide.
	4. Red-brown fumes in the cold or gentle warming	Bromine from bromate or bromide; chromyl chloride from a mixture of chloride and dichromate.
	5. Acid vapours and oily drops on the sides of the tube; a drop of water on the end of a glass rod turned turbid by the vapours	Fluoride or hexafluorosilicate.
	6. CO_2 and SO_2 evolved with charring of the sample	Tartrate
	7. Light brown acidic fumes on heating fairly strongly	Nitrogen dioxide from nitrate.
	8. CO_2 evolved	Probably oxalate.
	9. Gas giving positive $PdCl_2$ test for CO (below)	Carbon monoxide from formate, oxalate, citrate, tartrate, cyanide, hexacyanoferrate(II), hexacyanoferrate(III).
	10. Colourless gas; starch-iodide paper turned blue	Ozone from peroxodisulphate.
	11. Vigorous reaction; evil-smelling gas (mainly SO_2) evolved; sulphur deposited	Thiocyanate
	12. SO_2 evolved	Powerful reductants, e.g. phosphite, hypophosphite.

Test for Carbon Monoxide [4]

Cover the mouth of the tube with a strip of filter paper moistened with 2 drops of $PdCl_2$ solution. Heat for 1–2 minutes.

A grey-blue spot indicates carbon monoxide.

NOTE. The test is affected by sulphide, sulphite, bromide, iodide, and to a lesser extent, chloride ions. Sulphide and sulphite ions are best removed by treating the sample with 1 ml of $2M$ H_2SO_4 in the beaker and boiling to remove hydrogen sulphide or sulphur dioxide. The solution is concentrated to 2–3 drops and transferred to the centrifuge tube. The test is then carried out as above. In the presence of bromide or iodide ions, the sample is warmed gently with 1 ml of conc. sulphuric acid in the beaker and air from the teat pipette is blown through the mixture until the bromine or iodine has volatilized. The mixture is transferred to the centrifuge tube and heated again and the evolved gases are tested for carbon monoxide as above.

5.2.4 Tests on the Original Material for Interfering Anions

General Tests for Organic Anions [5]

A. Heat 3–5 mg of sample with 2 drops of 2% $K_2Cr_2O_7$ solution in $2M$ H_2SO_4 on a small watch-glass, almost to dryness. A deep green solution indicates the presence of an organic anion in the *absence* of inorganic reducing agents.

B. Treat about 20 mg of sample in a tube with 3 drops of $2M$ H_2SO_4 and warm. If CO_3^{2-} or OCN^- is present, boil out all the CO_2. Add about 20 mg of $K_2Cr_2O_7$, shake, add 3 drops of conc. H_2SO_4 and warm. Test for CO_2 (p. 158). CO_2 evolved shows an organic anion present. (If chloride ions are present in the sample, add 2 drops of Ag_2SO_4 solution before adding the $K_2Cr_2O_7$, to prevent evolution of chromyl chloride).

All organic anions considered herein except acetate give a positive reaction in these two tests.

Acetate

A. Add a few mg of sample to 3 drops of conc. H_3PO_4 in a semi-micro distillation apparatus.[†] Distil. Neutralize the distillate with $1M$ NH_3 (litmus paper) and mix 3 drops with $La(NO_3)_3$ solution. In another test-tube mix $0.01N$ iodine (2 drops) with $1M$ NH_3 (1 drop) and carefully add 2 drops of this, with minimal mixing, to the test solution. *A deep blue or grey-blue colour or precipitate at the interface, formed on standing, indicates acetate* [6]. Fluoride ions interfere. If they are present, carry out this test on the sodium carbonate extract after precipitating the fluoride with excess of calcium nitrate solution.

The test may also be carried out during the systematic anion separation.

B. Add 3 or 4 drops of conc. H_2SO_4 to 10–20 mg of the original sample and warm gently to drive off any gases. Add 2 drops of amyl alcohol and heat in a water-bath. *A fruity odour of amyl acetate indicates acetate.*

† Commercial apparatus is available but a modified teat pipette with teat removed is quite satisfactory. Bend the stem of a teat pipette through 90°. Insert the thin end into the centrifuge tube through a rubber bung. Distillate collects in the bulb of the teat pipette.

Borate

Thoroughly mix a few mg of the sample with an equivalent amount of CaF_2. Moisten with conc. H_2SO_4 and remove some of the paste on the loop of a platinum wire or on the end of a glass rod. Hold very close to the edge of the base of the burner flame *without* touching the flame. *A green flame indicates borate.* This test depends on the formation of volatile BF_3. Under the conditions used copper and barium compounds are not sufficiently volatile to colour the flame.

Fluoride [7, 8]

Place a few mg of the sample in a small depression in a lead plate (see p. 158). Add a little powdered silica or sodium silicate. Mix with a fine-pointed glass rod. Add 2 or 3 drops of conc. H_2SO_4 and cover with a square of thin perspex or other suitable plastic with a drop of water suspended on the underside. Let stand for about one minute. *A turbidity developing in the drop of water indicates fluoride.* The turbidity usually shows first round the edges of the drop.

Hexafluorosilicate [7, 8]

Apply the fluoride test without addition of the silicon compound. *A positive reaction indicates hexafluorosilicate.*

Silicate

Mix a small amount of the sample with NaF in the depression in the lead plate, using a fine-pointed glass rod. Carry out the fluoride test. *A positive reaction indicates silicate, provided hexafluorosilicate is absent.*

Oxidants

Warm a little of the sample with conc. HCl. Test for chlorine (yellow-green gas, smell, bleaches moist litmus paper). *Evolution of chlorine shows presence of a strong oxidizing agent.*

5.2.5 Preparation of the Sodium Carbonate Extract (Note 1)

The sodium carbonate extract is required both for the systematic analysis for anions and for some special tests.

Boil 0.1 g of the sample with 3.5 ml of 10% Na_2CO_3 solution (or 0.35 g of Na_2CO_3 and 3.5 ml of water) for at least 5 minutes, replacing any water lost by evaporation. During this process test for ammonia (below). Cool, centrifuge and reject the residue. Reserve 1 ml of the solution for the systematic anion scheme. Use the remainder for the tests described below (Note 2).

5.2.6 Tests on the Sodium Carbonate Extract

Ammonium

During the boiling in the preparation of the sodium carbonate extract smell the vapours evolved and test them with moist red litmus paper. *The evolution of a gas which turns the litmus paper blue and smells of ammonia indicates the presence of ammonium ions.*

Potassium

Prepare a solution of sodium hexanitrocobaltate(III) by mixing 12 drops of 1.5% $Co(NO_3)_2$ solution with 12 drops of 20% $NaNO_2$ solution and 2 drops of $4M$ acetic acid. Shake and allow to stand for 1 minute (Note 3). Add this mixture to 3 or 4 drops of the Na_2CO_3 extract, previously neutralized or made just acid with acetic acid. Shake and allow to stand for 1-2 minutes. Centrifuge. *A yellow precipitate indicates potassium.*

Formate [9] (Note 4)

To 4 drops of the Na_2CO_3 extract add a piece of litmus paper and then $4M$ acetic acid dropwise until the solution is just acidic. Shake to expel CO_2. Add a further drop of the acid. Remove the litmus paper. Add 1 drop of 5% $HgCl_2$ solution (Note 5). Bring to the boil and let stand for 1 minute. If a white precipitate forms, centrifuge and reject the solution. Add 2 or 3 drops of water to the precipitate, shake from side to side and centrifuge. Reject the solution. Add 1 drop of $4M$ NH_3 to the residue. *If the precipitate turns black, formate is indicated.*

Arsenite and Arsenate

To 3 or 4 drops of the Na_2CO_3 extract add a fragment of litmus paper and make just acid with $4M$ HCl. Add H_2S-acetone and warm. *A yellow precipitate indicates arsenite.* A slight orange precipitate may be obtained if antimony is present in the original sample. Centrifuge and reject the residue. Add some more H_2S-acetone to the solution to ensure that precipitation is complete. Add at least an equal volume of conc. HCl. Bring to the boil, cool and add H_2S-acetone. *A yellow precipitate indicates arsenate.* If no precipitate is obtained after the addition of H_2S, bring the solution to the boil again before assuming arsenate is absent.

Cyanide

Add a small crystal of $FeSO_4.7H_2O$ to 3 or 4 drops of the Na_2CO_3 extract and boil. Cool. Acidify with $4M$ HCl and add 1 drop of very dilute $FeCl_3$ solution. *A blue colour or precipitate indicates cyanide* (Note 6).

Nitrate and Nitrite

First establish the presence of nitrate and/or nitrite ions as follows. To 3 or 4 drops of the Na_2CO_3 extract add 0.5 ml of water and 5 or 6 drops of $4M$ NaOH. Add a few mg of Devarda's alloy and heat for 1-2 minutes, testing the vapours with moist red litmus paper. *NH_3 evolved indicates nitrate or nitrite ions* (Notes 7 and 8).

Nitrite

If nitrite ions are present, brown fumes will have been given off when the sample was treated with HCl in the preliminary tests.
A.   Acidify a few drops of the Na_2CO_3 extract with $2M$ H_2SO_4, and add 2 drops of $4M$ $FeSO_4$ solution. *A brown colour indicates nitrite.*

B.   Neutralize a few drops of the Na_2CO_3 extract with $4M$ acetic acid and add 0.5 ml of sulphanilic acid solution. Shake for at least 10 seconds and pour into 2 ml of a fresh solution of 1-naphthol in $2M$ NaOH. *A deep red colour indicates nitrite.*

NOTES
1.   Some insoluble substances are unaffected by boiling with sodium carbonate solution (see p. 129) although they contain ions which would normally be indicated by an examination of the sodium carbonate extract. Such materials must be dealt with by the 'insolubles' scheme, p. 213.
2.   If copper is present in the original sample, the sodium carbonate extract may be blue, owing to the formation of soluble carbonatocuprate(II) anions. Usually, copper has no effect on the tests for anions although copper sulphide may obscure the tests above for arsenite and arsenate. Reducing ions may produce a red precipitate of Cu_2O, however.
3.   This standing time is necessary for the oxidization of cobalt(II) to cobalt(III) by nitrous acid. Cobalt(III) then forms hexanitrocobaltate(III) ions.
4.   Arsenite, sulphide and thiosulphate ions interfere. If these are suspected to be present, treat the sodium carbonate extract as follows. To 4 drops of the extract, add 3 drops of 20 vol. H_2O_2 and boil for 30 seconds. Add a further 3 drops of 20 vol. H_2O_2 and boil for 1 minute. Cool and continue as in the original test.
5.   If iodide ions are present a reddish-yellow precipitate will be obtained on addition of mercury(II) chloride solution. If this occurs, centrifuge, then add a further drop of mercury(II) chloride solution. Continue until there is a slight excess of mercury(II) chloride. Remove the clear solution, bring to the boil and let stand as before.
6.   This test depends on the formation of hexacyanoferrate(II) ions. If hexacyanoferrate(II) is present in the original material, this cyanide test is invalid.
7.   Ammonia will also be evolved from hydroxylamine or hydrazine, if these are present.
8.   A positive confirmatory test for nitrite does not exclude the possibility of nitrate also being present.

5.2.7 Elementary Courses[†]

Apart from those anions indicated as possibly present by the cation analysis (arsenate, arsenite, chromate), it is doubtful whether anions other than carbonate, chloride, bromide, iodide, sulphate and nitrate would be included in an elementary scheme and it is generally sufficient to test directly for these anions. Mixtures of halides would not be expected at this level, but could be resolved by using the appropriate section of the systematic scheme (p. 194).

The following tests on the sodium carbonate extract may be used for elementary work. Take 3 or 4 drops for each test.

Sulphate. Acidify with a few drops of $4M$ HCl and heat. Add 2 drops of $BaCl_2$ solution. *A white precipitate indicates sulphate.*

Chloride. Acidify with a few drops of $4M$ HNO_3 and add 3 or 4 drops of $AgNO_3$ solution. *A white precipitate indicates chloride.* It is readily soluble in $2M$ NH_3.

†MAQA also publishes an elementary version of these separation tables.

Bromide. As for chloride. *A very pale yellow precipitate indicates bromide.* It is only partially soluble in $2M$ NH_3. Confirm on a fresh portion of the Na_2CO_3 extract as follows. Acidify with $2M$ H_2SO_4, add a small amount of PbO_2 and bring to the boil, holding a fluorescein test paper at the mouth of the tube. *A red colour indicates bromide.*

Iodide. As for chloride. *A pale yellow precipitate indicates iodide.* It is insoluble in ammonia. Confirm on a fresh portion of the Na_2CO_3 extract as follows. Acidify with $2M$ H_2SO_4 and add 1 or 2 drops of 2% $NaNO_2$ solution. *A brown colour or nearly black precipitate, dissolving in carbon tetrachloride to give a violet lower layer, indicates iodide.*

Chromate. Acidify 2 or 3 drops with $2M$ H_2SO_4 and cool. Add 0.5-1.0 ml of butanol–ether mixture and 1 or 2 drops of H_2O_2. Shake and allow the layers to separate. *A blue colour in the top layer indicates chromate.*

5.3 GROUP SEPARATION SCHEME FOR CATIONS

5.3.1 Preparation of Solution

First determine whether the sample dissolves in one of the solvents below. Take about 5 mg of sample in each instance and treat with a few drops of the solvent, testing in the order given until dissolution is achieved.

1. Water
2. $4M$ HCl
3. $4M$ HNO_3
4. Conc. HCl (Note 1)
5. Conc. HNO_3 (Note 2)

If the sample does not dissolve in cold solvent, bring to the boil and cool. When conc. HCl is used, some salts which are soluble in water (e.g. NaCl, $BaCl_2$) may not dissolve because of the high chloride ion concentration. In such instances dilution with water (0.5 ml) should yield a clear solution, and the observation can be used diagnostically. The sample may dissolve in acid but on dilution a precipitate may form as a result of the formation of insoluble basic salts by hydrolysis (e.g. BiOCl, SbOCl). Sufficient acid should be added to prevent precipitation, and note made of the probable presence of the cations responsible.

When a solvent is found, prepare a solution for analysis by dispensing about 30 mg of sample from the end of the spatula [10], and dissolving it in 2-3 ml of water or the minimum quantity of acid. Boil if necessary to remove any gases evolved. Cool, and dilute to 3 ml. It may be found that not all the components of a mixture dissolve in the solvents listed above. Any part of the sample which does not dissolve requires special treatment and should be analysed according to the scheme for analysing insoluble substances (p. 213).

NOTES

1. Concentrated hydrochloric acid dissolves some substances that are hardly affected by dilute acid (e.g. metallic tin, manganese dioxide). If manganese dioxide or permanganate is present, chlorine is evolved and the manganese(II) salt is formed. Similarly, dichromate is reduced to chromium(III) with evolution of chlorine. Thiosulphate gives a precipitate of sulphur with acids; if a Copper–Tin Group metal is present the sulphide is also formed, and barium salts are at least partially converted into insoluble barium sulphate. If thiosulphate is present, it should be dealt with as follows [11].

   Add 5 drops of 4M HCl (Note 3) to a 10-mg sample in a test-tube; warm and let stand for 1 minute. Boil for 10–15 seconds. Centrifuge.

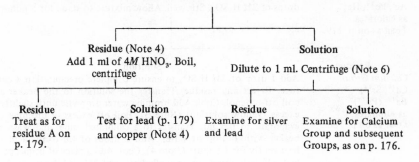

2. The use of concentrated nitric acid or *aqua regia* is rarely necessary. Dilute or concentrated nitric acid dissolves certain compounds of the Silver Group metals which obviously do not dissolve in hydrochloric acid; certain sulphides are also soluble. Nitric acid may precipitate metastannic acid from tin salts and hydrated antimony(V) oxide from antimony salts.

   If nitric acid (dilute or concentrated) is used to attack the sample, oxidation of mercury(I) to mercury(II), iron(II) to iron(III), bromide to bromine, and iodide to iodine takes place. If concentrated nitric acid is used, kinetically inert nitrato-complexes of chromium(III) may be formed, which do not react in the usual way, but trouble from this effect is seldom, if ever, encountered.

3. Dilute (4M) hydrochloric acid is preferred to concentrated because:
   (a) there is a cleaner separation of sulphur on centrifuging;
   (b) there is less oxidation of thiosulphate or sulphide to sulphate when oxidants are present; in the presence of nitrate and concentrated hydrochloric acid, a substantial amount of any barium present may be precipitated as the sulphate, and calcium and strontium sulphates may be precipitated to a lesser extent.

4. (a) The residue will also contain: (i) silver halides if present; (ii) barium sulphate if barium and nitrate are present; (iii) lead chloride if lead is present.
   If copper is present, a brownish-black precipitate of copper(II) sulphide, which remains after boiling, is sometimes formed. Copper may be confirmed by extracting the residue, after centrifuging, with 4M nitric acid and adding ammonia solution, to give a deep blue colour.
   (b) If arsenic is present a yellow sulphide precipitate forms.
   Arsenic can be confirmed by treating the residue with dilute potassium hydroxide solution and testing in the usual manner (p. 185). Arsenic will again precipitate in the Copper–Tin Group.
   However, if both copper and arsenic are present with thiosulphate, normally neither is precipitated as sulphide by treatment with 4M hydrochloric acid.

5. Mixtures containing mercury(I) or mercury(II) with thiosulphate decompose. Silver chloride and bromide are slightly soluble in 4M hydrochloric acid, and lead chloride is appreciably soluble; dilution causes reprecipitation of the material dissolved.

5.3.2 Separation of the Cations into Groups

Take 1 ml of the prepared solution (p. 174), which should contain about 10 mg of sample. If conc. HNO_2 (or *aqua regia* p. 175) has been used to dissolve the sample, first evaporate to dryness and take up the residue in 1 ml of water. Add 4M HCl until precipitation is complete (Note 1). Centrifuge.

Residue	Solution
Silver Group Ag⁺, Pb²⁺, Hg₂²⁺ as chlorides. Treat as on p. 179.	Add an equal volume of ethanol and boil. (Note 2). Cool, and add 5 drops of 2M H_2SO_4. Stir well. Allow mixture to stand for 5 minutes.

Residue	Solution
Calcium Group Ca²⁺, Sr²⁺, Ba²⁺, (Pb²⁺) as sulphates. Treat as on p. 180.	Add 1 drop of 2M H_2SO_4 to ensure complete precipitation. Centrifuge. Reject any residue. Transfer the solution to the beaker and boil off ethanol. Cool. Add bromine water dropwise until slightly in excess (yellow solution) (Note 3). Boil off excess of bromine and evaporate to 0.5 ml. Transfer to a calibrated tube, and wash the beaker with 1 drop of water. Add 2 drops of 1M oxalic acid and heat gently for 1 minute (Note 4). Cool. Add a piece of litmus paper and 4M NH_3 dropwise until alkaline. Add 2M HCl dropwise (Note 5), shaking well after each addition (Note 6), until the litmus turns red. Add 0.25 ml in excess (Note 7) and heat gently for 3 minutes. Dilute to 1.5 ml with water. *If arsenic is absent,* add 6–8 drops of H_2S-acetone [12], mix well and warm gently. Centrifuge and add 2 or 3 drops more to ensure complete precipitation. *If arsenic is present* add 4 drops of 10% NH_4I solution [13] (Note 8). Boil. Add 6–8 drops of H_2S-acetone and boil again. Cool and centrifuge.

Residue	Solution
Copper–Tin Group Cu²⁺, Hg²⁺, Bi³⁺, Cd²⁺, As(III), Sb(III), Sn(IV) as sulphides. Treat as on p. 182.	Evaporate to near dryness. Remove interfering anions, except phosphate (Note 9). Dilute with 10–15 drops of water. Test for phosphate (Note 10), and remove if present (Note 9). Add 2 drops of conc. HNO_3 and boil (Note 11). Add 5 drops of 20% NH_4Cl (Note 12), a piece of litmus paper, and 4M NH_3 dropwise, until just in excess (Note 13). Centrifuge.

Residue	Solution
Iron Group Fe³⁺, Al³⁺, Cr³⁺ as hydrated oxides. Treat as on p. 187.	Add H_2S-acetone. Centrifuge.

Residue	Solution
Zinc Group Zn²⁺, Mn²⁺, Co²⁺, Ni²⁺ as sulphides. Treat as on p. 188.	*Magnesium Group* Examine for Mg²⁺, Na⁺, Li⁺ (p. 190).

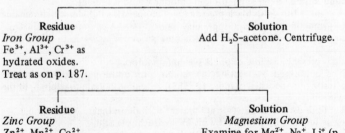

NOTES
1. A white precipitate at this stage is not always a Silver Group chloride (p. 99).
    (a) Borates, in concentrated solution, may yield a white precipitate of boric acid.
        Only partial precipitation will occur, because the acid is appreciably soluble. Its
        presence (which should already have been established) does not affect the identi-
        fication of Silver Group chlorides. The amount remaining in solution can be
        removed after precipitation of the Copper–Tin Group.
    (b) Silicates may give a gelatinous precipitate of hydrated silica. Because silicate
        should already have been identified in the preliminary tests, this should be expec-
        ted. It should not interfere unduly with the Silver Group separation. Only partial
        precipitation will occur and it is essential to remove the remainder after precipita-
        tion of the Copper–Tin Group, otherwise it will be precipitated in the Iron Group
        on addition of ammonium chloride and may be confused with aluminium. It is
        easily removed by evaporating to dryness with dilute hydrochloric acid 2 or 3
        times. The hydrated silica is converted into a granular form, which can be centri-
        fuged and separated after treatment of the residue with dilute hydrochloric acid.
    (c) Certain organic acids (e.g. benzoic, salicylic) may precipitate when the solution is
        acidified. If their presence is suspected from the preliminary tests, treat the
        original solution with 2 or 3 drops of dilute nitric acid, centrifuge and separate
        the precipitate and proceed with the analysis on the solution.
    (d) Certain chlorides (e.g. sodium chloride, barium chloride) in very concentrated
        solution may precipitate owing to the common ion effect. They will dissolve on
        dilution with water. Compounds of bismuth and antimony may precipitate as the
        oxide chlorides if the solution is insufficiently acidic.
    (e) A coloured precipitate of the appropriate metal sulphide may form if thio-salts
        of arsenic, antimony or tin are present.
2. Any lead not precipitated in the Silver Group should precipitate when ethanol is added.
    On boiling with ethanol, permanganate and dichromate ions are reduced to man-
    ganese(II) and chromium(III) ions; it is necessary to carry out the reduction before
    adding dilute sulphuric acid, to avoid co-precipitation with barium sulphate.
3. Bromine water oxidizes tin(II) to tin(IV) and ensures precipitation of tin(IV) sul-
    phide with hydrogen sulphide. Tin(IV) sulphide dissolves readily in alkaline solutions.
    Oxidation with bromine is unnecessary, however, if the chloroacetate separation is to
    be used later, because tin(II) sulphide dissolves readily in the alkaline chloroacetate
    solution. Arsenite and iron(II) are also oxidized by bromine water to arsenate and
    iron(III) respectively.
4. Too much oxalic acid must be avoided, as this may prevent the subsequent precipita-
    tion of tin and antimony. The oxalic acid prevents the precipitation of tin(IV) hy-
    droxide on adding ammonia solution. This precipitate is sometimes difficult to dissolve
    in dilute hydrochloric acid.
5. Prepare about 2 ml of $2M$ hydrochloric acid by mixing an equal number of drops of
    $4M$ hydrochloric acid and water. Some of this acid will be required later in the separa-
    tion of the Silver Group (p. 179) and of the Tin Group sulphides (p. 184).
6. If iron is present, the precipitate of hydrated iron(III) oxide dissolves very slowly and
    if the tube is not shaken well to accelerate dissolution, too much acid may be added.
    The precipitate obscures the colour change of litmus but this can be overcome by
    tilting the tube so that the paper sticks to the upper wall. After the addition of each
    drop of acid, the solution is allowed to come into contact with the litmus paper and
    the colour is observed.
7. For the precipitation of the Copper–Tin Group sulphides, the acidity should be between
    0.3 and $0.5M$; if the hydrochloric acid concentration is too high, certain sulphides,
    e.g. cadmium sulphide, tin(IV) sulphide, are incompletely precipitated or not pre-
    cipitated at all.
        It may not be possible to add exactly 0.25 ml of $2M$ hydrochloric acid with any
    given pipette but the following approximations can be used. It must be borne in
    mind that the drop of $2M$ HCl which turns the litmus red may amount to one complete
    drop in excess or it may be almost completely neutralized. This is taken into account
    in the range of acidities listed below.

Drop volume of pipette	*No. of drops required*	*Range of acid concentration*
0.05 ml	5	0.33 – 0.40M
0.06 ml	4	0.32 – 0.39M
0.07 ml	4	0.37 – 0.43M
0.08 ml	3	0.32 – 0.44M
0.09 ml	3	0.36 – 0.45M

8. Hydrogen sulphide does not precipitate arsenic from an arsenate solution which is only slightly acidic. However, if hydrogen sulphide is present for a long period, reduction to arsenic(III) takes place followed by precipitation of arsenic(III) sulphide; hence the use of ammonium iodide for the reduction of arsenic(V).

A reď or yellow precipitate of mercury(II) iodide may appear if mercury(II) is present. It should be dissolved in excess of ammonium iodide solution and a further 3 drops of ammonium iodide solution added.

A slight white precipitate of copper(I) iodide may form if copper(II) is present; this can be ignored because hydrogen sulphide will convert it into copper(I) sulphide.

9. *Removal of interfering anions* (p. 127).

(a) *Organic acids.* Repeated evaporation with conc. nitric acid eliminates organic material, complete removal usually being indicated by the disappearance of the carbonaceous residue. Residual amounts of salicylic acid (Note 1) form picric acid under these conditions, which may be hazardous. The addition of 2 or 3 drops of dilute sulphuric acid before the evaporation with nitric acid prevents the residue from drying out before decomposition of the organic matter is complete.

If organic acids are present as well as other interfering anions, the organic acids should be removed first. Remember that oxalate has been added, and must be removed at this stage.

(b) *Silicate.* This has been dealt with in Note 1, above. Prolonged baking of the residue may convert oxides of iron(III), aluminium(III), chromium(III) or magnesium(II) into an insoluble form and a fusion will then be necessary to dissolve them.

(c) *Phosphate.* Dilute to about 1 ml with water. Add 4 drops of 4M nitric acid and boil. Add about 0.2 g of solid ammonium chloride and dissolve it. Add zirconium nitrate solution drop by drop until precipitation is complete. Only a small excess of reagent should be present and it is preferable to add a few drops at a time, centrifuging after each addition. Heat just to boiling and centrifuge. Wash the precipitate with a few drops of water and add the washings to the previously separated liquid. Discard the residue of zirconium phosphate and proceed to the Iron Group with the solution. Excess of zirconium precipitates in the Iron Group as white zirconium hydroxide and separates with iron but does not interfere with the confirmatory test for iron. If the presence of phosphate has been confirmed, but no precipitate is formed on the addition of zirconium nitrate, add 1 drop of 10% diammonium hydrogen phosphate solution to overcome supersaturation and continue as described above.

(d) *Fluoride.* Add 5 drops of conc. hydrochloric acid, evaporate nearly to dryness, and repeat this treatment three more times.

(e) *Borate.* If an Iron Group metal is present borate will co-precipitate in the Iron Group and it will be necessary to remove the borate, after precipitation of the Copper–Tin Group metals, by volatilization as ethyl borate. Add 1 ml of conc. hydrochloric acid and 1 ml of ethanol. Evaporate nearly to dryness and repeat the evaporation with hydrochloric acid and ethanol.

10. *Phosphate test.* Add, with shaking, 6 drops of molybdate reagent B to 2 drops of molybdate reagent A (p. 235). Add 3 or 4 drops of water and 3 drops of test solution. Warm gently. *A yellow precipitate indicates phosphate.* If iodide is present, the precipitate may be pale green, but boiling the precipitate with 2 drops of conc. nitric acid will turn it yellow.

11. Iron(II) hydroxide is not precipitated completely by ammonia so nitric acid is added to oxidize iron(II) to iron(III). If iron(III) was present originally it would have been at least partially reduced by hydrogen sulphide to iron(II) (with deposition of sulphur).
12. In the presence of excess of ammonium chloride there should theoretically be no precipitation of manganese, but the possibility should never be overlooked. A certain amount of manganese(II) hydroxide is always present, and may be oxidized by air to insoluble hydrated manganese(IV) oxide, which is precipitated with the Iron Group metals. This is generally overcome by (i) adding an excess of ammonium chloride; (ii) boiling to expel dissolved air before adding ammonia solution; (iii) centrifuging as quickly as possible.
13. Ensure that the ammonia solution is not just in the upper layer, by stirring well with a rod. Ammonia may be present in the vapour phase even though excess is not present in solution. The surface above the liquid should be blown clear of vapour before the smell is tested and relied on (p. 154).
14. A brown solution contains colloidal nickel sulphide. Coagulate by making just acidic with 4M acetic acid and boiling.

5.3.3 Silver Group
Ag^+, Pb^{2+}, Hg_2^{2+}

Wash the precipitate once with 0.5 ml of 2M HCl (see Note 5 above). Add 2 ml of water, boil and stir. Centrifuge and separate while hot.

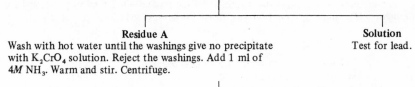

Residue A	Solution
Wash with hot water until the washings give no precipitate with K$_2$CrO$_4$ solution. Reject the washings. Add 1 ml of 4M NH$_3$. Warm and stir. Centrifuge.	Test for lead.

Residue	Solution
Black indicates mercury(I). Wash with water. Test for mercury(I).	Test for silver.

Confirmatory Tests

One confirmatory test is usually sufficient to identify a cation.

Lead

A. To part of the hot solution add 1 drop of 2% KI solution. *A yellow precipitate crystallizing in silky plates indicates lead.*
B. To part of the hot solution add 3 or 4 drops of 2M ammonium acetate and 1 drop of 5% K$_2$CrO$_4$ solution. *A yellow precipitate indicates lead.*

Mercury

Add to the residue 3 or 4 drops of conc. HCl and 2 or 3 drops of bromine water. Heat to dissolve and boil off excess of Br$_2$. Dilute to about 1 ml with water:

A. To part of the solution add 1 drop of 5% SnCl$_2$ solution. *A white or grey precipitate indicates mercury.*

B.   To part of the solution add $4M$ NH_3 till just alkaline, then $4M$ HNO_3 till just acid. Add 2 or 3 drops of 0.03% p-dimethylaminobenzylidenerhodanine solution. *A red colour indicates mercury.*

Silver

A.   To part of the solution add 1 drop of 2% KI solution. *A yellow precipitate indicates silver.*

B.   Acidify the remainder of the solution with $4M$ HNO_3. *A white precipitate indicates silver.* Centrifuge and wash the precipitate once with 1 ml of water. Suspend the precipitate in about 0.5 ml of water and add 2 or 3 drops of 0.03% p-dimethylaminobenzylidenerhodanine solution. *A red colour indicates silver.*

5.3.4 Calcium Group [14, 15]

$$Ca^{2+}, Sr^{2+}, Ba^{2+}, (Pb^{2+})$$

Wash the residue by stirring with 1 ml of ethanol to which 1 drop of $2M$ H_2SO_4 has been added. Centrifuge. Discard the washings. To the residue add 1.5 ml of water, and stir thoroughly whilst bringing to the boil. Centrifuge the hot mixture.

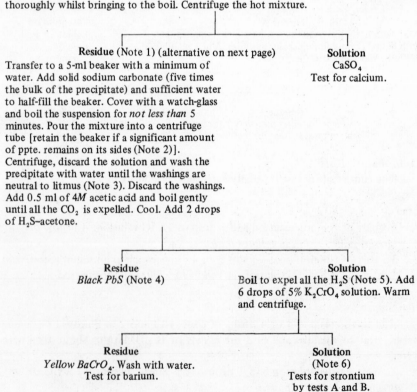

Residue (Note 1) (alternative on next page)
Transfer to a 5-ml beaker with a minimum of water. Add solid sodium carbonate (five times the bulk of the precipitate) and sufficient water to half-fill the beaker. Cover with a watch-glass and boil the suspension for *not less than* 5 minutes. Pour the mixture into a centrifuge tube [retain the beaker if a significant amount of ppte. remains on its sides (Note 2)]. Centrifuge, discard the solution and wash the precipitate with water until the washings are neutral to litmus (Note 3). Discard the washings. Add 0.5 ml of $4M$ acetic acid and boil gently until all the CO_2 is expelled. Cool. Add 2 drops of H_2S–acetone.

Solution
$CaSO_4$
Test for calcium.

Residue
Black PbS (Note 4)

Solution
Boil to expel all the H_2S (Note 5). Add 6 drops of 5% K_2CrO_4 solution. Warm and centrifuge.

Residue
Yellow $BaCrO_4$. Wash with water.
Test for barium.

Solution
(Note 6)
Tests for strontium by tests A and B.

Alternative separation

Residue

Transfer to a 5-ml beaker with 1.0 ml of ammoniacal 0.5M EDTA. Warm to dissolve (Note 7). Add
0.5 ml of 10% K_2CrO_4 solution, warm, and add
slowly, with stirring, 0.5 ml of 1.2M $MgCl_2$. Stir
for 1 minute, transfer to a centrifuge tube,
centrifuge.

Residue	**Solution**
Yellow *BaCrO₄*. Wash with water.	Transfer to a 5-ml beaker and add
Test for barium.	0.5 ml of 1M $ZnCl_2$. Warm gently
	and add 1 ml of ethanol. Transfer
	to a centrifuge tube. Centrifuge.
	A yellow residue indicates stron-
	tium. Wash with 50% ethanol-
	water. Test for strontium by
	tests C and D.

Confirmatory Tests

Calcium

A. To a portion of the solution add 2 drops of saturated ammonium oxalate
    solution. Allow to stand for 5 minutes, scratching the inside of the tube
    with a glass rod if necessary. *A white precipitate indicates calcium.*

B. To 3 drops of the test solution add 4 drops of 1% glyoxal-bis(2-hydroxyanil)
    solution, 1 drop of 2M NaOH and 1 drop of 10% Na_2CO_3 solution. Shake
    the mixture with 4 drops of chloroform. *A red chloroform layer indicates*
    calcium.

C. To the remainder of the test solution add a little solid NH_4Cl and 5 drops
    of 5% $K_4Fe(CN)_6$ solution. *A white precipitate indicates calcium.*

Strontium

Carbonate method

Divide the solution into halves.

A. To one half add 5 drops of saturated $CaSO_4$ solution. *A white precipitate,*
    sometimes slow to form, indicates strontium.

B. To the other half add 5 drops of 5% K_2CrO_4 solution and evaporate to 0.5
    ml. Add 4M NH_3 until the orange solution turns yellow, and add a further
    drop of K_2CrO_4 solution. Dilute with water to 0.75 ml. If no precipitate
    appears, add 0.25 ml of ethanol (Note 8). *A yellow precipitate, sometimes*
    slow to form, indicates strontium.

EDTA method

C. Dissolve the residue in 1 drop of 4M acetic acid and 5 drops of water. Place
    1 drop of the solution on filter paper and add 1 drop of fresh sodium
    rhodizonate solution. A red-brown spot is formed. Fume it over conc. NH_3
    solution. *If the spot remains, strontium is indicated.*

D. To the remainder of the solution add 1 drop of $2M$ H_2SO_4 and 4 drops of ethanol. Centrifuge, reject the solution, moisten the residue with conc. HCl and apply the flame test. *A crimson flame indicates strontium.*

Barium

A. Boil the washed precipitate with 5 drops of conc. HCl. Cool, centrifuge and separate the $BaCl_2$ crystals. Wash the crystals with 3 drops of conc. HCl. Dissolve the residue in 5 drops of water and apply the flame test. *An apple-green flame indicates barium.*

B. To the aqueous solution obtained for the flame test add 3 drops of sodium rhodizonate solution. *A bright red colour or precipitate indicates barium.*

NOTES
1. If the residue is small compared to the original Group precipitate, it is possible that only calcium sulphate is present and the residue should be further tested with 1 ml of water.
2. Wash the residue with water until the washings are neutral to litmus (Note 3), dissolve it in 3 drops of $4M$ acetic acid, and add to the main solution obtained with 0.5 ml of acetic acid. Wash the beaker with a further 3 drops of acetic acid and again transfer to the main solution.
3. The residue must be washed free from sulphate, otherwise the metal sulphate will reprecipitate on treatment with acetic acid.
4. Any lead not precipitated in the Silver Group is precipitated as lead sulphate in the Calcium Group. If lead was not found in the Silver Group, the precipitates should be dissolved in hot $4M$ nitric acid and a confirmatory test for lead (p. 179) applied.
5. Lead acetate paper can be used to detect hydrogen sulphide.
6. The solution will also contain any calcium sulphate not extracted initially with water.
7. A slight turbidity may remain if barium sulphate is present. The presence of lead can be confirmed at this stage. One drop of the solution is spotted onto a filter paper and treated with 1 drop of H_2S–acetone. *A brownish-black spot indicates lead.*
8. The volume of ethanol must be measured accurately. Too much results in the precipitation of calcium if this ion is present in sufficient amount.

5.3.5 Parting of the Copper–Tin Group

(An alternative scheme is given on p. 186)

Wash the precipitate with water. Reject the washings. Add to the precipitate 2 ml of $0.5M$ KOH (Note). Warm gently and stir for 2 minutes. Centrifuge.

Residue	Solution
Copper Group	Tin Group
Mercury(II), bismuth, cadmium, copper(II) as sulphides.	Arsenic, antimony, tin as thio-salts.
Treat as on p. 183.	Treat as on p. 184.

NOTE
The use of $0.5M$ potassium hydroxide for the separation of the Copper Group from the Tin Group is based on the work of James and Woodward (*Analyst*, 1955, 80, 825) and others. It appears that no reagent of this type can effectively separate the two groups under all conditions [16], and if mercury and Tin Group metals are present together, mercury should be sought in both the Copper and the Tin Groups.

5.3.6 Copper Group

Hg^{2+}, Bi^{3+}, Cd^{2+}, Cu^{2+}

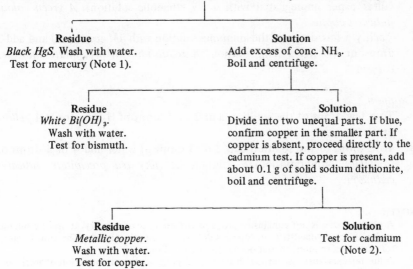

Wash the residue *twice* with 1 ml of water. Add 1 ml of $4M$ HNO_3 and boil. Centrifuge.

Residue
Black HgS. Wash with water.
Test for mercury (Note 1).

Solution
Add excess of conc. NH_3.
Boil and centrifuge.

Residue
White Bi(OH)$_3$.
Wash with water.
Test for bismuth.

Solution
Divide into two unequal parts. If blue,
confirm copper in the smaller part. If
copper is absent, proceed directly to the
cadmium test. If copper is present, add
about 0.1 g of solid sodium dithionite,
boil and centrifuge.

Residue
Metallic copper.
Wash with water.
Test for copper.

Solution
Test for cadmium
(Note 2).

Confirmatory Tests
Mercury

Dissolve HgS by boiling with 2 or 3 drops of conc. HCl and 2 or 3 drops of bromine water. Boil off excess of Br_2. Dilute to about 1 ml with water.

A. To a few drops of the solution add 2 drops of 5% $SnCl_2$ solution. *A white precipitate turning grey indicates mercury.*

B. To a few drops of the solution add 2 or 3 drops of 0.03% *p*-dimethyl-aminobenzylidenerhodanine solution and 5 drops of 50% sodium acetate solution. *A pink colour indicates mercury.*

Bismuth

A. Dissolve a portion of the precipitate in a few drops of $4M$ HNO_3 and add 2 drops of 10% thiourea solution. *A yellow colour indicates bismuth.*

B. Dissolve a portion of the precipitate in the minimum volume of $4M$ HCl and pour the solution into sodium stannite solution, freshly prepared by adding $4M$ NaOH to 5 ml of 5% $SnCl_2$ solution until the precipitate just dissolves. *A black precipitate indicates bismuth.*

C. Dissolve a portion of the precipitate in the minimum volume of $4M$ HCl. Add 2 drops of 1% dimethylglyoxime solution and make ammoniacal. *A yellow precipitate indicates bismuth.*

Copper

A. Dissolve the residue in 0.5-1 ml of 4*M* HNO₃. To the solution add excess of conc. NH₃ solution. *A deep blue solution indicates copper.*

B. Spot one drop of the blue solution from A onto α-benzoin oxime test paper (filter paper impregnated with a 5% ethanolic solution). *A green colour indicates copper.*

C. Acidify a few drops of the ammonia solution with 4*M* acetic acid and add 2 drops of 5% K₄Fe(CN)₆ solution. *A reddish-brown precipitate indicates copper.*

Cadmium

A. To a few drops of the solution add 2 or 3 drops of H₂S–acetone. *A yellow precipitate indicates cadmium.*

B. To 2 drops of the solution add 2 or 3 drops of water and 4 or 5 drops of iron(II)/2,2'-bipyridyl/iodide solution. *A silky red precipitate indicates cadmium.*

NOTES

1. A black residue is not conclusive proof of the presence of mercury. It may be sulphur enclosing small quantities of copper sulphide or bismuth sulphide. The confirmatory test for mercury must, therefore, always be carried out.

2. Some mercury may be carried through as a result of the solvent action of nitric and hydrochloric acids on mercury(II) sulphide if the precipitate was not originally washed free from hydrochloric acid. Although some would be precipitated by the ammonia solution, the mercury compound is soluble in ammonium salt solutions and a little passes into solution. However, it will not interfere with the cadmium tests after the separation with dithionite because it will separate as the metal with copper.

5.3.7 Tin Group (Note 1)

As(III), Sb(III), Sn(IV)

Just acidify the KOH extract with 2*M* HCl and add 4–6 drops of H₂S–acetone. Centrifuge and discard the solution. Wash the residue with water. Add 2 ml of 4*M* HCl. Boil for 5–10 seconds. Cool. Add a few drops of H₂S–acetone. Centrifuge.

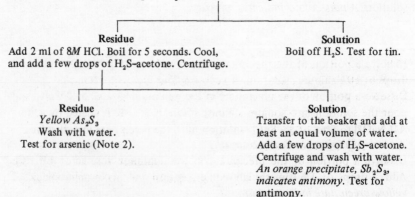

Residue
Add 2 ml of 8*M* HCl. Boil for 5 seconds. Cool, and add a few drops of H₂S–acetone. Centrifuge.

Solution
Boil off H₂S. Test for tin.

Residue
Yellow As₂S₃
Wash with water.
Test for arsenic (Note 2).

Solution
Transfer to the beaker and add at least an equal volume of water. Add a few drops of H₂S–acetone. Centrifuge and wash with water. *An orange precipitate, Sb₂S₃, indicates antimony.* Test for antimony.

Confirmatory Tests

Arsenic

A. Dissolve a part of the precipitate in the minimum volume of $4M$ NH_3 with a few drops of 20 vol. H_2O_2. Dilute to 1 ml and boil off excess of ammonia. Cool. Add 5 drops of 1% $AgNO_3$ solution. *A red-brown precipitate indicates arsenic.*

B. Dissolve a part of the precipitate by boiling with conc. HNO_3 and add excess of molybdate reagent A (p. 235). Boil. *A yellow precipitate indicates arsenic.*

C. Dissolve a part of the precipitate in $4M$ NH_3 + H_2O_2 as in A. Add a few drops of magnesia mixture. Stir occasionally and let stand for 2–3 minutes. *A white precipitate indicates arsenic.*

Tin

A [17]. To 4 or 5 drops of solution add 4 drops of 30% tartaric acid solution, 5 drops of water and 6 drops of 1% *N*-benzoylphenylhydroxylamine in 50% acetic acid. *A white precipitate indicates tin.*

B. To the remainder of the solution add a little iron wire and boil for 10–15 seconds, with brisk evolution of H_2. Cool. Add 2 drops of this solution to 1 drop of 0.25% cacotheline solution. *A violet colour indicates tin.* The oxidation state of tin in the original material should be determined. The following tests on the original sample are used to detect tin(II).

  (i) *Sulphide and thiosulphate absent.* Treat 10 mg of the sample with 1 ml of cold $4M$ HCl. Centrifuge if necessary. Add 2 drops of this solution to 1 drop of cacotheline solution. *A violet colour indicates tin(II).*

  (ii) *Sulphide present.* To 10 mg of sample add 1 ml of $4M$ HCl and boil for 30 seconds. Cool and centrifuge if necessary. Apply the cacotheline test to the solution as in (i).

  (iii) *Thiosulphate present.* Obtain a solution as in (i). Add 5 drops of $0.5M$ $BaCl_2$. Stir for 1 minute to hasten precipitation. Centrifuge and apply the cacotheline test as in (i) to the clear solution.

C. Tin(IV) may be identified by dissolving a part of the original sample in $4M$ HCl, centrifuging if necessary, and applying Test A.

Antimony

Dissolve the orange precipitate of antimony sulphide in sufficient conc. HCl diluted with an equal volume of water and boil to expel H_2S. To two drops of this solution add 2 drops of 2% $NaNO_2$ solution and 0.5 ml of 0.01% Rhodamine B solution. *A violet colour or precipitate indicates antimony.*

NOTES
1. This separation depends entirely on the use of both 4*M* and 8*M* hydrochloric acid to dissolve selectively tin(IV) sulphide and antimony sulphide, respectively, from the mixed sulphides. The acids *must* be of the correct strength; if the conditions are adhered to, a very clean separation will be obtained.
2. The presence of arsenic indicates the presence of arsenite and/or arsenate in the original mixture.
    If mercury passes through to the Tin Group it will remain with the arsenic after the Group separation. Its presence should be obvious because of the black colour of the precipitate. To separate arsenic and mercury, wash the residue with water and boil with 1 ml of saturated $(NH_4)_2CO_3$ solution. Centrifuge.

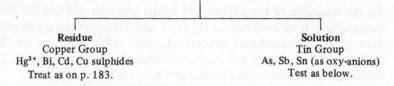

Residue	Solution
Black HgS. Wash with water. Dissolve in a few drops of conc. HCl and bromine water and test for mercury.	Add two drops of 20 vol. H_2O_2. Boil and test for arsenic.

5.3.8 Parting of the Copper–Tin Group (Alternative Procedure)

Wash the precipitate with water. Reject the washings. Add to the precipitate 2 ml of sodium chloroacetate solution. Heat at 60–80° for 3 minutes (Note). Centrifuge.

Residue	Solution
Copper Group	Tin Group
Hg^{2+}, Bi, Cd, Cu sulphides	As, Sb, Sn (as oxy-anions)
Treat as on p. 183.	Test as below.

Tests for Tin Group Metals

The following tests may be carried out on 1 ml of the solution in a centrifuge tube.

Add 2 or 3 aluminium turnings. Cover the mouth of the tube with a filter paper impregnated with $AgNO_3$ solution. Warm the solution gently until it boils. *A grey or black stain of elementary silver indicates arsenic. A black precipitate formed in the tube after the arsenic test indicates antimony.*

Centrifuge, and retain the supernatant liquid to test for tin. Wash the precipitate well with water, drain it well, and dissolve it in *aqua regia*. Dilute the solution and apply the antimony test (p. 185).

Acidify the solution from the antimony precipitation with 4*M* HCl and add a little iron wire. Boil for 10–15 seconds. Cool. Add 4 drops of this solution to 1 drop of cacotheline solution. *A violet colour indicates tin.*

NOTE
Heating is essential to convert the sulphide ions into thiodiglycollate ions. Insufficient heating merely achieves the effect of other alkaline solutions and mercury may be found in both Groups. Completeness of reaction can be tested by acidifying a few drops of the solution. A sulphide precipitate will form if reaction is incomplete.

5.3.9 Iron Group (Note 1)
Fe^{3+}, Al^{3+}, Cr^{3+}

Wash the precipitate with water. Reject the washings. Add 2 or 3 drops of $4M$ NaOH (Note 2) and 4 or 5 drops of water. Boil and centrifuge.

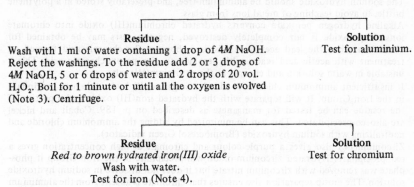

Residue
Wash with 1 ml of water containing 1 drop of $4M$ NaOH. Reject the washings. To the residue add 2 or 3 drops of $4M$ NaOH, 5 or 6 drops of water and 2 drops of 20 vol. H_2O_2. Boil for 1 minute or until all the oxygen is evolved (Note 3). Centrifuge.

Solution
Test for aluminium.

Residue
Red to brown hydrated iron(III) oxide
Wash with water.
Test for iron (Note 4).

Solution
Test for chromium

Confirmatory Tests
Iron

A. Dissolve a portion of the precipitate in a few drops of $4M$ HCl and add 1 drop of $2M$ NH_4SCN. *A blood-red colour indicates iron.*

B. Dissolve as in test A and add 1 drop of $K_4Fe(CN)_6$ solution. *A deep blue precipitate indicates iron.*

The oxidation state of iron in the original material should be determined. Dissolve a small portion of the original sample in $4M$ HCl (if the sample does not dissolve completely, warm and centrifuge, and reject the residue). Divide the solution into two parts.

(i) To a few drops of solution add 1 drop of 5% $K_4Fe(CN)_6$ solution. *A blue colour or precipitate indicates iron(III).*

(ii) To a few drops of solution add 1 drop of fresh $K_3Fe(CN)_6$ solution. *A blue colour or precipitate indicates iron(II). A brown solution or precipitate indicates iron(III).*

Aluminium

To 2 or 3 drops of the test solution add 1 drop of phenolphthalein solution and neutralize carefully with $1M$ HCl, *adding not more than 1 drop in excess.* Add 5 drops of $2M$ ammonium acetate, followed by 5 drops of 0.1% Solochrome Cyanine solution. *An intense purple colour indicates aluminium* (Note 5) [18].

Chromium

A. Acidify with $4M$ acetic acid and add 2 drops of 10% lead acetate solution. *A yellow precipitate indicates chromium.*

B. Acidify with $2M$ H_2SO_4 *Cool thoroughly under running water.* Add 1 drop of 20 vol. H_2O_2 and 0.5 ml of butanol/ether mixture and shake the tube from side to side. *A blue top layer indicates chromium.*

NOTES
1. Small amounts of Zinc Group metals may co-precipitate in this Group. They will not interfere with the tests, nor will sufficient have been lost to prevent their later detection. Sodium hydroxide and hydrogen peroxide may produce insoluble black cobalt(III) oxide if cobalt is present, which will darken the colour of hydrated iron(III) oxide.
2. The sodium hydroxide should be aluminium-free, and preferably stored in a polythene bottle, to avoid leaching of metal ions from glass.
3. Alkaline hydrogen peroxide converts hydrated chromium(III) oxide into chromate ions. If peroxide is not completely destroyed, negative tests may be obtained for chromium when the lead acetate test is used, because the chromic acid formed on treatment with acetic acid reacts with peroxide to give peroxochromic acid. This is unstable in warm solution and changes to the chromium(III) salt.
4. If insufficient ammonium chloride is added, manganese(II) hydroxide may precipitate in the Iron Group. It will separate with the hydrated iron(III) oxide and a portion of the residue can be tested for manganese as described on p. 189. Cobalt and nickel are also co-precipitated. This can be minimized by adding the ammonium chloride and neutralizing with sodium hydroxide (Bromocresol Green indicator).
5. Zirconium(IV) also gives a purple colour and chromate at high concentration gives a red precipitate. Hydrated zirconium oxide will precipitate in the Iron Group if phosphate was removed with zirconium nitrate but it does not dissolve in sodium hydroxide solution. The group separation also ensures that chromate is absent from the aluminium test solution.

5.3.10 Zinc Group [19]

$$Zn^{2+}, Mn^{2+}, Ni^{2+}, Co^{2+}$$

Wash the mixed sulphides with 1 ml of water. Reject the washings. Add 4 or 5 drops of conc. HCl and 2 or 3 drops of conc. HNO_3 (Note 1). Warm to dissolve and transfer to the beaker or crucible. Evaporate to dryness, in the fume cupboard (Note 2). To the residue add 1–1.5 ml of $1M$ acetic acid and heat to boiling. Transfer to a centrifuge tube and add H_2S–acetone.

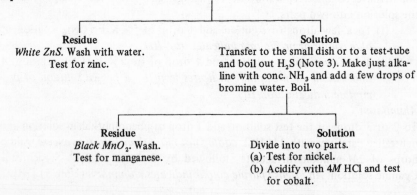

Residue
White ZnS. Wash with water.
Test for zinc.

Solution
Transfer to the small dish or to a test-tube
and boil out H_2S (Note 3). Make just alkaline with conc. NH_3 and add a few drops of bromine water. Boil.

Residue
Black MnO_2. Wash.
Test for manganese.

Solution
Divide into two parts.
(a) Test for nickel.
(b) Acidify with $4M$ HCl and test for cobalt.

Confirmatory Tests

Zinc

Dissolve the residue in the minimum amount of $2M$ H_2SO_4. Dilute with a little water. Boil out H_2S. Cool and divide into two portions.

A. Add 1 drop of 5% $K_4Fe(CN)_6$ solution. *A white precipitate indicates zinc.*

B. Concentrate to 0.5 ml if necessary. Add 2 drops of 0.1% $CuSO_4$ solution and 2 drops of ammonium tetrathiocyanatomercurate(II) solution. Stir. *A violet-black precipitate indicates zinc* (Note 4).

Manganese

A. Dissolve part of the residue in 0.5 ml of $4M$ HNO_3 and 2 drops of 20 vol. H_2O_2. Boil till all the oxygen is evolved. Cool. Add 0.25 g of sodium bismuthate. Centrifuge. *A purple solution indicates manganese.*

B. To the rest of the residue add a few drops of $2M$ H_2SO_4 with 1 drop of dilute $AgNO_3$ solution and a little solid ammonium peroxodisulphate. Boil. *A purple colour indicates manganese.*

Nickel

Add 2 drops of 1% dimethylglyoxime solution followed by $4M$ NH_3 until alkaline. Warm. *A red precipitate indicates nickel.*

Cobalt

A. To 3 or 4 drops of the solution add 1 drop of 5% $SnCl_2$ solution, about 0.25 g of solid NH_4SCN and 0.5-1 ml of butanol/ether mixture. Shake the tube from side to side. *A blue colour in the top layer indicates cobalt.*

B. *If nickel is absent.* To a portion of the solution add $4M$ NaOH till alkaline, then just acidify with $4M$ HCl. Add 2 or 3 drops of 0.5% nitroso-R salt solution followed by 0.5 ml of $4M$ sodium acetate. *A red colour indicates cobalt.*

    If nickel is present. Precipitate the nickel as the dimethylglyoximate (above) and reject the residue. Make the solution just acid with $4M$ HCl, add 2 or 3 drops of 0.5% nitroso-R salt solution and 0.5 ml of $4M$ sodium acetate. Bring to the boil and allow to stand for a few minutes. *A red-brown colour indicates cobalt.*

NOTES
1. Instead of concentrated nitric acid, 3 drops of 10% sodium hypochlorite solution can be used.
2. Evaporation must be to complete dryness in order to remove oxidizing agents; otherwise some sulphur may precipitate when hydrogen sulphide is added. If the residue is overheated, and blackened, it may not dissolve completely in dilute acetic acid.
3. If hydrogen sulphide is not boiled out, addition of ammonia solution will precipitate cobalt and nickel sulphides.
4. Scratching the inside of the tube facilitates precipitation.

5.3.11 Magnesium Group [20, 21]
Mg^{2+}, Na^+, Li^+

Evaporate the solution to about 0.5 ml. Transfer the solution to a centrifuge tube and add 2 drops of $4M$ NH$_3$ followed by 1 ml of 3% 8-hydroxyquinoline solution in chloroform. Shake well and centrifuge. Remove the upper aqueous layer with a teat pipette and transfer to a separate tube.

Yellow chloroform layer	Aqueous layer
Wash with 0.5 ml of water and reject the aqueous layer. Transfer to the crucible and evaporate the chloroform. Heat the residue strongly until all the 8-hydroxy-quinoline has vaporized and a grey-white residue remains (Note 1). Cool, and test for magnesium.	Repeat the extraction with 1 ml of 8-hydroxyquinoline solution and discard the chloroform layer. Divide the solution into two parts. (i) Test for sodium. (ii) Add 4 or 5 drops of $4M$ KOH and boil down to ca. 0.25 ml. Ensure that the test solution is alkaline and that all the NH$_3$ has vola-tilized. Test for lithium.

Confirmatory Tests

Magnesium

To the residue add 4 or 5 drops of $4M$ HCl. Transfer the solution to a tube and centrifuge. Reject any residue. To the solution add 2 drops of 0.1% Titan Yellow solution followed by $4M$ NaOH until the solution is alkaline. *A turbid red solution yielding a red precipitate on standing or centrifuging indicates magnesium* (Note 2). Carry out a blank test on the acid used to dissolve the oxide residue.

Sodium

Evaporate to dryness. Dissolve in 1 drop of $4M$ acetic acid. Add at least 5 times the volume of copper uranyl acetate solution (Note 3). Mix well, leave for a few minutes and centrifuge. *A pale yellow crystalline precipitate indicates sodium.*

Lithium

To the solution add $4M$ HCl dropwise until *just* acid, then add 1 drop of $4M$ KOH to make alkaline. Add 2 drops of saturated NaCl solution and twice the volume of iron(III) potassium periodate solution (Note 4). Mix and leave for a few minutes. *A white precipitate indicates lithium.*

NOTES
1. There is no loss of magnesium at this stage through volatilization of its 8-hydroxy-quinoline complex. Metals which might have escaped precipitation in the earlier groups of the analytical scheme (e.g. calcium) will be present in the residue.
2. The Titan Yellow test is highly sensitive and specific for magnesium. However, doubt may exist as to the exact colour of the magnesium hydroxide–Titan Yellow lake, and it is recommended that the test be done at the same time on (i) distilled water, (ii) a solution containing 0.1–0.2 mg of magnesium per ml, and (iii) the test solution, and the results compared.
3. Copper uranyl acetate is insensitive to lithium whereas the more common zinc uranyl acetate gives a precipitate with both sodium and lithium.
4. A saturated solution of sodium chloride does not react at room temperature with the reagent, but at 90–100°C the sodium complex is precipitated. It is permissible to warm to 50°C for 15–20 seconds but further heating causes precipitation of the sodium compound, in the absence of lithium.

5.4 GROUP SEPARATION SCHEME FOR ANIONS [22, 23]
The presence of the following anions will already have been established (pp. 170-173).

acetate[†]	fluoride[†]	phosphate[†]
arsenate[†]	formate[†]	silicate[†]
arsenite[†]	hexafluorosilicate	sulphite
borate[†]	nitrate	thiosulphate
cyanide [†]	nitrite	

Evidence for the presence of the following anions may have been obtained in the preliminary examination:

bromate[†]	chloride[†]	permanganate
bromide[†]	chromate[†]	peroxodisulphate
carbonate	dithionite	sulphide[†]
chlorate[†]	iodide[†]	thiocyanate

It is essential to confirm the presence of all anions which are *not* accommodated in the systematic scheme, by using appropriate tests, such as those given on pp. 197-202.

It should be remembered, however, that certain anions are incompatible and therefore would not be present together in solution, e.g. arsenite and bromate, sulphite and chromate.

5.4.1 Removal of Interfering Anions
Hypochlorite, hexafluorosilicate and peroxodisulphate ions are destroyed during the extraction with sodium carbonate, and therefore are not included in the systematic scheme of analysis.

Cyanide, thiosulphate and permanganate ions interfere in the systematic scheme and must be destroyed. Cyanide and thiosulphate ions are destroyed by adding 5 or 6 drops of 20 vol. hydrogen peroxide to the sodium carbonate extract and boiling for 5 minutes. Cyanide is converted into cyanate which does not interfere, thiosulphate is converted into sulphate. Permanganate is destroyed by adding 20 vol. hydrogen peroxide dropwise until the sodium carbonate extract is decolourized. The manganese(IV) oxide is separated by centrifugation and the solution is boiled for 5 minutes to destroy the excess of hydrogen peroxide. In this process, certain other ions are affected:

$$
\begin{aligned}
\text{sulphite} &\longrightarrow \text{sulphate} \\
\text{sulphide} &\longrightarrow \text{sulphate} \\
\text{hexacyanoferrate(III)} &\longrightarrow \text{hexacyanoferrate(II)} \\
\text{arsenite} &\longrightarrow \text{arsenate} \\
\text{periodate} &\longrightarrow \text{iodate}
\end{aligned}
$$

[†]Anions included in the systematic scheme, where they may be identified without any preliminary information.

Hence, if sulphate, arsenate, iodate or hexacyanoferrate(II) is found after this treatment, it is necessary to test the original material or the sodium carbonate extract for the presence of the ion found and also for the ions from which it could have been derived.

5.4.2 Separation Scheme for Anions

Prepare some silver ammine reagent by placing 0.1 ml of silver ammine solution A (p. 238) in a tube and adding solution B dropwise (Note 1), until the precipitate which forms just disappears. Add 1 ml of the Na_2CO_3 extract (p. 171) and centrifuge.

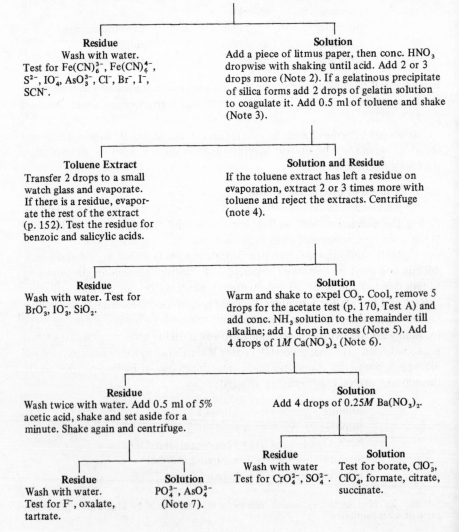

Residue
Wash with water.
Test for $Fe(CN)_6^{3-}$, $Fe(CN)_6^{4-}$, S^{2-}, IO_4^-, AsO_3^{3-}, Cl^-, Br^-, I^-, SCN^-.

Solution
Add a piece of litmus paper, then conc. HNO_3 dropwise with shaking until acid. Add 2 or 3 drops more (Note 2). If a gelatinous precipitate of silica forms add 2 drops of gelatin solution to coagulate it. Add 0.5 ml of toluene and shake (Note 3).

Toluene Extract
Transfer 2 drops to a small watch glass and evaporate. If there is a residue, evaporate the rest of the extract (p. 152). Test the residue for benzoic and salicylic acids.

Solution and Residue
If the toluene extract has left a residue on evaporation, extract 2 or 3 times more with toluene and reject the extracts. Centrifuge (note 4).

Residue
Wash with water. Test for BrO_3^-, IO_3^-, SiO_2.

Solution
Warm and shake to expel CO_2. Cool, remove 5 drops for the acetate test (p. 170, Test A) and add conc. NH_3 solution to the remainder till alkaline; add 1 drop in excess (Note 5). Add 4 drops of $1M$ $Ca(NO_3)_2$ (Note 6).

Residue
Wash twice with water. Add 0.5 ml of 5% acetic acid, shake and set aside for a minute. Shake again and centrifuge.

Solution
Add 4 drops of $0.25M$ $Ba(NO_3)_2$.

Residue
Wash with water.
Test for F^-, oxalate, tartrate.

Solution
PO_4^{3-}, AsO_4^{3-}
(Note 7).

Residue
Wash with water
Test for CrO_4^{2-}, SO_4^{2-}.

Solution
Test for borate, ClO_3^-, ClO_4^-, formate, citrate, succinate.

NOTES
1. *Only 4-6 drops of solution B should be required;* if much more than this is necessary, prepare and use a new solution B.
2. Silver oxalate may precipitate unless the solution is strongly acidic. Ignore any intermediate precipitate which redissolves.
3. If no organic matter was found during the preliminary examination or the systematic examination for cations, nó extraction need be carried out.
4. If chromate ions are present in large amounts they may precipitate here. This yellow precipitate is easily recognized and it does not interfere with the tests for iodate and bromate ions, even if it contains some precipitated silica.
5. A precipitate may form at this stage. It may be further silica or hydrated oxides of aluminium and tin compounds in the sodium carbonate extract. The precipitate should be centrifuged and discarded.
6. Calcium tartrate precipitates slowly. The solution should be cooled in water and the inside wall of the tube scratched with a glass rod to promote precipitation. The tube should be allowed to stand for 5 minutes.
7. Phosphate and arsenate should have been detected during the cation separation and further tests should be unnecessary.

Tests on the Residue from the Silver Ammine Precipitation

If the precipitate is white, silver sulphide (black), silver hexacyanoferrate(III) (orange), silver arsenite (pale yellow), silver periodate (dark brown) and silver iodide (pale yellow) are likely to be absent. It is advisable, however, always to test for iodide and arsenite, for the pale colours may not show sufficiently in the presence of white silver salts.

Arsenic need not be tested for if it was not found during the examination for cations, unless further confirmation is required. *Sulphide* should have been detected in the preliminary tests and need not be tested for here unless further confirmation is required.

Test the precipitate in the following order.

A. *Hexacyanoferrate(II)*

Place a portion of the precipitate on a filter paper, spot with conc. HCl and 2% $FeCl_3$ solution. *A blue colour indicates hexacyanoferrate(II).*

B. *Hexacyanoferrate(III)*

Place a portion of the precipitate on a filter paper, spot with 1 drop of conc. HCl and a drop of freshly prepared $FeSO_4$ solution. *A blue colour indicates hexacyanoferrate(III).*

C. *Sulphide*

Place a portion of the precipitate on a filter paper, spot with 1 or 2 drops of 10% $Na_2S_2O_3$ solution. *A black stain remaining indicates sulphide.*

D. *Other anions*

Extract the rest of the precipitate with 50% v/v HNO_3 (if periodate is present the brown silver salt turns white before dissolution). Centrifuge.

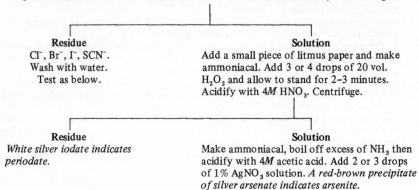

Residue	Solution
Cl^-, Br^-, I^-, SCN^-.	Add a small piece of litmus paper and make
Wash with water.	ammoniacal. Add 3 or 4 drops of 20 vol.
Test as below.	H_2O_2 and allow to stand for 2–3 minutes.
	Acidify with $4M$ HNO_3. Centrifuge.

Residue	Solution
White silver iodate indicates	Make ammoniacal, boil off excess of NH_3 then
periodate.	acidify with $4M$ acetic acid. Add 2 or 3 drops
	of 1% $AgNO_3$ solution. *A red-brown precipitate*
	of silver arsenate indicates arsenite.

Tests for Halides and Thiocyanate [24]

To the halide-thiocyanate precipitate, add 10 drops of conc. NH_3 solution and 3 drops of yellow ammonium sulphide solution. Stir the mixture with a glass rod and warm until the black silver sulphide coagulates. Centrifuge and reject the precipitate. Transfer the solution to the beaker and boil to expel NH_3 and to decompose the ammonium sulphide. When the solution becomes cloudy, add 5 or 6 drops of $4M$ $HClO_4$ (Note 1) and continue heating until H_2S is completely removed. Carry out the following tests.

A. Place 1 drop on filter paper; add 1 drop of *iodide-free* starch solution (Note 2) and 1 drop of 20% $NaNO_2$ solution. *A blue colour indicates iodide.*

B. Add 5 drops of $1M$ $Pb(NO_3)_2$ to the solution in the beaker and transfer to a tube. Centrifuge (Note 3). Transfer the solution to a clean beaker and place 1 drop on a filter paper. Add 1 drop of 1% $NH_4Fe(SO_4)_2$ solution. *A red colour indicates thiocyanate.*

C. If thiocyanate or iodide is present, add 3 or 4 drops of 20% $NaNO_2$ solution (Note 4) and boil until no more brown fumes are evolved (if thiocyanate and iodide are absent, proceed directly with the test for bromide). Remove the beaker from the heating block, allow to cool somewhat and add 2 or 3 drops of $4M$ $HClO_4$ (Note 5). Add a spatula-end of PbO_2 and quickly cover with a watch-glass with a strip of moistened fluorescein paper on the underside. Boil gently for a few seconds and rinse the paper with water. *A red colour indicates bromide.*

D. If bromide is present, boil the uncovered contents of the beaker for 30 seconds to expel the remainder of the bromine. Allow to stand for 30 seconds and decant the solution into a tube. Centrifuge if necessary. Add 1% $AgNO_3$ solution. *A white precipitate indicates chloride.*

NOTES
1. The metathesis of the mixed silver salts results in the release of chloride, bromide, iodide, thiocyanate and hexacyanoferrate(II) ions. Hexacyanoferrate(III) originally present is reduced by the sulphide in alkaline solution to hexacyanoferrate(II). Dilute perchloric acid is used to acidify the solution to avoid the oxidation of hexacyanoferrate(II) to hexacyanoferrate(III) which occurs when nitric acid is used. The perchlorate ion does not interfere with the subsequent reactions.
2. Freshly prepared starch solution is used. Alternatively, and more conveniently, the commercial starch indicator 'Thyodene' dissolved in sufficient water is suitable for the test.
3. Lead nitrate precipitates all the hexacyanoferrate(II), most of the iodide, and some of the thiocyanate ions. Sulphur arising from the decomposition of the ammonium sulphide is also present at this stage and is removed with the precipitated lead salts by centrifuging. The formation of a bright yellow precipitate at this stage confirms the presence of iodide.
4. Nitrous acid oxidizes the small amount of iodide remaining in solution and the iodine is expelled by boiling. Thiocyanate is decomposed with the formation of hydrogen cyanide, nitric oxide and sulphate. A precipitate of lead sulphate appears if thiocyanate was present.
5. It is important that the solution should remain acidic during the treatment with nitrite and during the subsequent tests. If insufficient acid is present during the nitrite treatment, brown fumes of nitrogen dioxide appear on the addition of more acid and these must be boiled out before the bromine test is made.

Tests for Salicylate and Benzoate [25]

A. Dissolve a small portion of the residue in the minimum amount of water. Add 1 drop of 1% $NH_4Fe(SO_4)_2$ solution. *An intense purple colour indicates salicylate.*

B. Dissolve the remainder of the residue in the minimum amount of water. Acidify with $2M$ H_2SO_4. Add bromine water dropwise until precipitation is complete and a permanent yellow colour forms. Shake with 1 ml of dilute Na_2CO_3 solution. Centrifuge and reject any precipitate (Note 1). Add 3 drops of 20 vol. H_2O_2 to the solution. Acidify with $2M$ H_2SO_4, and extract into toluene as before. Evaporate the toluene. Place the residue in a *dry* centrifuge tube, add 1 ml of Le Rosen's reagent and warm gently (Note 2). *A brown-red colour indicates benzoate.*

NOTES
1. Salicylate is precipitated as tribromophenol and separated, to prevent interference in the benzoate test.
2. It is recommended that a blank be run alongside the benzoate test.

Tests for Bromate, Iodate and Silica

Treat the washed precipitate with a few drops of $4M$ NH_3, centrifuge any silica, warm the remaining solution, and add 3 drops of yellow ammonium sulphide. Boil for a few seconds to reduce the halates completely to halides; centrifuge and reject the precipitated silver sulphide. Transfer the solution to the beaker and boil to expel ammonia and to decompose the ammonium sulphide. Test for bromide and iodide as described above. *Positive tests indicate bromate and iodate.*

The first residue from the test may be tested for silica, by the lead plate test (p. 158), but its presence should already have been established (p. 171).

Tests for Fluoride, Oxalate and Tartrate
It should not be necessary to test for fluoride because it should have been identified earlier (p. 171). If confirmation is required use the lead plate test (p. 158). Test the precipitate as follows.

A. Transfer part of the precipitate to the beaker, add 4 drops of conc. H_3PO_4 and a spatula-end of diphenylamine. Boil for 1-2 minutes to remove water, allow to cool for 30 seconds, add 0.5 ml of ethanol and stir. *A deep blue colour indicates oxalate* [26].

B. Dissolve the remainder of the precipitate by adding a few drops of water followed by dropwise addition of $4M$ HNO_3. When a clear solution is obtained, add 2 or 3 drops more of $4M$ HNO_3. Add an equal volume of silver periodate reagent and warm. Do not boil. *A white precipitate indicates tartrate.*

Tests for Sulphate and Chromate
Add 0.5 ml of $4M$ HCl to the precipitate, warm and centrifuge. *A white precipitate remaining indicates sulphate.* Transfer the solution to another centrifuge tube, and add 0.5 ml of water. Cool thoroughly. Add 1 drop of 20 vol. H_2O_2 and 0.5 ml of butanol/ether mixture. Shake the tube from side to side. *A blue colour in the upper layer indicates chromate.*

Tests for Anions Remaining in Solution
The solution may contain borate, acetate, chlorate, citrate, formate, perchlorate and succinate ions, and also sodium, barium, calcium and traces of bromate ions. Borate and acetate should have been found previously (pp. 170, 171) and need not be tested for again. Test the solution as follows.

A. To 3 drops of the test solution, add $2M$ H_2SO_4 until precipitation of $BaSO_4$ is complete. Centrifuge and discard the precipitate. Take 1 drop of the solution, add 4 drops of saturated $ZnSO_4$ solution and 1 drop of 0.03% Methylene Blue solution. *A purplish-red colour indicates perchlorate* [27].

B. Take one third of the test solution, make $4M$ in HNO_3, add 50 mg of $NaNO_2$ and boil. Cool, add 1 drop of silver ammine solution A. *A white precipitate indicates chlorate* (Note 1) [27].

C. To a portion of the test solution add 4 drops of silver ammine solution A (Note 2). Centrifuge immediately.

Residue	**Solution**
Wash twice with water. Dissolve in 5 drops of $4M$ NH_3. If formate is present heat to boiling, centrifuge, and reject any precipitate. Divide the solution into two portions and test for citrate and succinate, as follows.	Heat just to boiling. *An intense blackening indicates formate.*

(i) Add to the test solution 3 drops of Denigès reagent. Centrifuge and discard the precipitate. Add 2 drops of $0.02M$ $KMnO_4$ and warm. *Decolourization of the $KMnO_4$, followed by formation of a heavy white precipitate, indicates citrate.*

(ii) Add 4 drops of conc. HCl, centrifuge and discard the precipitate (Note 3). Evaporate to dryness. Add 6 drops of conc. NH_3 solution. Transfer to a centrifuge tube and carefully evaporate to dryness. Avoid excessive heating. Add 2 spatula-tip quantities of zinc dust, and mix well with the residue. Heat strongly for a few seconds, then test the gases evolved with a piece of filter paper soaked in *p*-dimethylaminobenzaldehyde solution and inserted into the gas-testing funnel. Heat strongly. *The rapid production of a deep purple colour indicates succinate* (Note 4) [25].

NOTES
1. If bromate ions are present, traces will be found in the test solution, and will be reduced to bromide. The precipitate obtained with silver nitrate should therefore be tested for bromide and chloride ions (p. 194).
2. Some borate ions may precipitate, but do not interfere with the citrate and succinate tests.
3. If citrate ions are present, they should first be eliminated as follows. Add to the remaining test solution 3 spatula-tip quantities of powdered potassium permanganate and bring the solution *just* to the boil. Add conc. HCl drop by drop until all the permanganate is reduced. Centrifuge, and proceed with the test as above.
4. If the fumes decolourize the paper, insert a fresh piece and continue as before.

5.4.3 Tests for Certain Combinations of Anions

Nitrate in the Presence of Nitrite

In a fresh portion of the sodium carbonate extract, destroy nitrite by boiling the *neutralized* solution with an equal volume of $4M$ NH_4Cl. To a few few drops of the nitrite-free solution, add three times the volume of acetate buffer and 10–20 mg of zinc dust. Shake for 15–20 seconds and centrifuge. Test the solution for nitrite as on p. 172. Alternatively use Devarda's alloy (p. 172) on the nitrite-free solution, but ensure that all ammonia is first boiled out.

Carbonate, Hydrogen Carbonate and Cyanate (Note 1) (see p. 137) [28]

Shake 50 mg of sample with 1 ml of water and centrifuge.

A. *Residue.* Wash with a little water and add $4M$ acetic acid (Note 2); *effervescence and lime water turned milky by the gas indicates carbonate.* If there is no effervescence, carbonate is absent. If present, destroy carbonate completely with $4M$ acetic acid. Add an excess of $4M$ HCl (Note 3). *Further evolution of CO_2 indicates cyanate.* Make alkaline with $4M$ NaOH and boil. *NH_3 evolved confirms cyanate.*

B. *Solution.* Divide into three equal portions, and test as follows.

(i) Add 4M HCl (Note 2); *effervescence and lime water turned milky by the gas indicates carbonate, hydrogen carbonate and/or cyanate.*

(ii) Add 1 drop of phenolphthalein solution; *a red colour confirms carbonate.* If the solution remains colourless or shows not more than a very faint pink colour, *carbonate is absent.* Add saturated BaCl$_2$ solution dropwise until the colour is discharged and add 3 drops in excess. *Slow effervescence in the cold or on gently warming confirms hydrogen carbonate.*

(iii) Add 1 drop of phenolphthalein solution, make alkaline with 4M NaOH and boil. Ammonia evolved confirms the presence of an ammonium salt, in which case boil for 1 minute to remove all the ammonia. Make acidic with 4M HCl; *evolution of CO$_2$ indicates carbonate and/or cyanate.* Make alkaline again with 4M NaOH and boil. *NH$_3$ evolved confirms cyanate.*

NOTES

1. The equilibrium between carbonate and hydrogen carbonate ions will be disturbed by the hydrogen ions liberated when salts of strong acids and weak bases are dissolved in water. Solutions containing such salts and a carbonate will contain appreciable amounts of hydrogen carbonate ions.

2. If sulphite ions were detected in the preliminary test with dilute acid, add 2 drops of 20 vol. hydrogen peroxide before the addition of acid.

3. The acetate exerts a buffering action on the hydrolysis of cyanate. It is advisable to count the drops of acetic acid used and to add the same number of drops of hydrochloric acid plus 3 or 4 more.

Sulphide, Sulphite and Thiosulphate

Sulphide

Dilute 6 drops of the Na$_2$CO$_3$ extract to 1 ml with water and add 10% CdSO$_4$ solution dropwise. *A yellow precipitate indicates sulphide.* Continue the additions, centrifuging between each, until the new precipitate produced is pure white. Avoid a large excess of the reagent (Note 1). Retain the solution for the tests below. To the residue add 4M HCl dropwise and test the evolved gas with lead acetate paper. *Blackening of the paper confirms sulphide.*

Thiosulphate

Test a small portion of the solution from the sulphide test for complete removal of sulphide by adding 1 drop of lead acetate solution (Note 2). Take 6 drops of the remaining solution, make just acid with 2M H$_2$SO$_4$ and add 0.5 ml of 1% AgNO$_3$ solution (Note 3). *A cream-yellow precipitate turning brown, then black on gentle warming, indicates thiosulphate.*

Sulphite

To the remainder of the solution from the sulphide test add 1 ml of saturated HgCl$_2$ solution and centrifuge for *1 minute* (Note 4). Add a further drop of $HgCl_2$ solution to ensure the complete removal of thiosulphate and carbonate ions. To the solution add 1 ml of $0.5M$ $BaCl_2$ (Note 5) and centrifuge. Wash the precipitate and treat with a few drops of $4M$ HCl; warm and test for SO_2 by suspending a drop of starch–iodine solution in the vapour. *Decolourization of the starch–iodine indicates sulphite.*

NOTES
1. This procedure completely removes sulphide and some carbonate ions; a large excess of cadmium should not be present, otherwise cadmium sulphite may precipitate.
2. A white precipitate of lead carbonate should be formed. If black lead sulphide is precipitated, a further drop of cadmium sulphate solution must be added to the main solution which should then be centrifuged.
3. Many anions can be precipitated at this stage (e.g. halides) but the colour change on gentle warming is characteristic of thiosulphate.
4. Any cloudiness remaining is probably sulphur formed in the reaction between thiosulphate ions and mercury(II) chloride. This can be used as a further test for thiosulphate. After centrifuging, treat the residue with $4M$ HCl. A yellow suspension is obtained if thiosulphate ions are present but the solution remains clear if they are absent.
5. The purpose of the barium precipitation is:
    (a) to concentrate the sulphite ions as a precipitate, from which sulphur dioxide is subsequently liberated,
    (b) to separate the sulphite ions from nitrite ions which, if present, interfere with the detection of sulphur dioxide.

Thiosulphate, Sulphite, Dithionate and Polythionate [29]

Dissolve 0.1 g of sample in 5 ml of water.

Thiosulphate

To 0.2 ml of test solution add 1 drop of $4M$ HCl and allow to stand. *A white turbidity indicates thiosulphate.*

Dithionate and Thiosulphate or Sulphite

To 0.2 ml of test solution add 1 drop of $4M$ HCl and immediately add $0.01M$ I_2 dropwise. *Disappearance of the iodine colour indicates thiosulphate or sulphite.*

Continue adding I_2 solution until there is one drop in excess, giving a pale brown solution. Transfer to a beaker and add an equal volume of conc. HCl. Add $0.01M$ KIO_3 dropwise until any brown colour that develops is rendered very pale yellow or colourless. Heat to boiling. *A golden brown colour indicates dithionate.* (If the solution is pale yellow, a portion should be left unboiled, so that a comparison may be made with the boiled solution).

Sulphite

To 1 ml of test solution add 4 ml of pH 4 acetate buffer solution followed by 4 drops of $0.5M$ $BaCl_2$. Divide the solution into two portions. To one add $0.01M$ I_2 dropwise until in excess. *A white turbidity (in comparison with the untreated portion of the solution) indicates sulphite.*

Polythionates (thiosulphate and sulphite absent).

To 0.2 ml of test solution add 0.2 ml of conc. HCl, followed by 3 drops of 0.01M KIO$_3$. *A brown colour indicates a polythionate* (not dithionate).

Chromate and Dichromate [30]

Take 50 mg of sample, shake with 1 ml of water and centrifuge. Subject the residue to sodium carbonate metalthesis (p. 171) and test the filtrate for chromate. The detection of chromate indicates the presence of an insoluble chromate in the sample. Test the solution as follows.

Chromate

Evaporate 3 drops of the test solution to dryness in a crucible and allow to cool. Add 2 drops of nickel–dimethylglyoxime solution and allow to stand for 2 minutes. *A red precipitate indicates chromate* (Note 1).

Dichromate

Evaporate 3 drops of the test solution to dryness in a crucible and allow to cool. Add 1 drop of iodide–iodate solution and 1 drop of 1% starch solution and allow to stand for 5 minutes. *A blue colour indicates dichromate* (Note 2).

NOTES
1. If only a small amount of chromate is suspected, a comparison test should be made on a pure dichromate solution. A trace of red precipitate sometimes forms with the latter, but it does not adhere to the surface of the crucible when water is added and the crucible is emptied, as it does when chromate is present.
2. If only a small amount of dichromate is suspected, a comparison test should be made on a pure chromate solution. A slight blue colour which fades on standing is given by chromate alone, whereas the blue colour deepens on standing in the presence of dichromate.

Phosphite and Hypophosphite [31, 32]

To 0.5 ml of the test solution (Note 1) add a slight excess of saturated bromine water and boil until the yellow colour disappears. Cool, then add 0.1 g of NH$_4$NO$_3$ and 0.5 ml of 10% ammonium molybdate solution. Warm. *A yellow precipitate indicates the presence of phosphate* or *phosphite and/or hypophosphite.*

If the test is positive, take a further 0.5 ml of the test solution (Note 1) and add 0.5 ml of 10% ammonium molybdate solution followed by 2 drops of 2M H$_2$SO$_4$. Warm but *do not boil. If the solution gradually turns blue, hypophosphite is indicated. If the solution remains colourless or becomes faintly yellow (Note 2) phosphite is indicated. If a bright yellow precipitate is rapidly produced, phosphate is indicated.*

NOTES
1. About 10 mg of the test substance should be dissolved and the final volume should be 1 ml. If the test sample is insoluble in *warm* water, it should be dissolved either in *warm*

4M hydrochloric or 2M sulphuric acid or in *warm* 4M sodium hydroxide; if the sample is dissolved in acid, the solution should be neutralized with 4M sodium hydroxide and the insoluble metal hydroxide removed by centrifuging; if dissolved in alkali, the metal (lead) must be precipitated as its sulphate with 2M sulphuric acid and removed by centrifuging.

When preparing the test solution *do not boil* it, otherwise phosphite and hypophosphite ions will be converted into phosphate ions.
2.   A faint cream or yellow precipitate may be produced on prolonged standing.

Hypochlorite and Chlorine Water

Make the test solution slightly acid with 2M H_2SO_4 (Note), then shake it with mercury. *A brown precipitate soluble in 4M HCl indicates hypochlorite; a white precipitate insoluble in 4M HCl indicates chlorine water.*

NOTE
Alkaline hypochlorites give an alkaline reaction with litmus paper.

5.5 TESTS FOR PARTICULAR SUBSTANCES

Hydrazine and Hydroxylamine

Dissolve a few crystals of the material under test in 3 drops of water and add 1 drop of conc. H_2SO_4. Shake. If a white crystalline solid begins to separate, set aside for about 1 minute. Centrifuge (Note 1).

Make the centrifuged solution alkaline with 9M NH_3 solution, then just acidic with 2M H_2SO_4. Add 5 drops of salicylaldehyde solution and shake. If a cloudiness appears, set aside for 5 minutes. Centrifuge for at least 1 minute (Note 2). *A white precipitate confirms hydrazine.*

Add 3 drops of copper(II) acetate solution to the centrifuged solution and shake. If a cloudiness appears, set aside for about 1 minute. Centrifuge. *A yellow-green precipitate confirms hydroxylamine.*

NOTES
1.   Most of the hydrazine is removed as its sparingly soluble sulphate. The residual hydrazine is then precipitated with salicylaldehyde on standing for 5 minutes.
2.   It is extremely difficult to obtain a completely clear solution on centrifuging the precipitate.

Sulphamates

Treat 2 drops of the test solution with 2 drops of $BaCl_2$–$NaNO_2$ solution and then add 2 drops of 4M HCl. *A white precipitate (or turbidity in the case of small amounts) indicates a sulphamate* (Note).

NOTE
When carbonate, sulphate or other anions which give insoluble barium salts are present, the test solution should be treated first with 0.5M barium chloride, the resultant precipitate removed by centrifuging, and the solution then tested as above. Peroxodisulphate (see below) will give a precipitate on warming the solution.

Peroxodisulphates

To 2 drops of the test solution add 1 drop of $4M$ HNO_3 and 2 or 3 drops of $0.25M$ $Ba(NO_3)_2$, centrifuge and discard any precipitate. Boil the solution for about 2 minutes. *A white precipitate indicates a peroxodisulphate.* Confirmation is given by the evolution of ozone on heating an aqueous solution of the material (p. 167).

Hydrogen Peroxide

To 2 drops of the test solution add 1 drop of $2M$ H_2SO_4 and 2 drops of $Ti(SO_4)_2$ solution. *A yellow-orange colour indicates hydrogen peroxide.*

5.6 THE ANALYSIS OF MIXTURES ALSO CONTAINING LESS COMMON ELEMENTS

The foregoing schemes for the systematic examination of the commoner cations and anions have been extended to include the following less common elements: beryllium, cerium, molybdenum, selenium, tellurium, thallium, thorium, titanium, tungsten, uranium, vanadium and zirconium.

When these elements are present, the earlier instructions must be modified as described below. Most of the common and less common elements in admixture may be separated and detected by using the extensions given below, but detailed provision for *every* possible mixture has not yet been made.

The following additions and amendments to the schemes should be made if less common elements are present.

5.6.1 Preliminary Dry Tests
Flame test (p. 165)

Thallium compounds impart an intense transient green colour to the flame, which is quite distinct from the green colour given by copper and barium salts. Some selenium compounds give a bright blue flame, and some tellurium compounds a blue-green flame.

5.6.2 Preliminary Examination for Anions and Potassium
Special Tests on the Sodium Carbonate Extract (p. 171)

Potassium. The portion of the Na_2CO_3 extract taken for this test should be made acidic with acetic acid and any precipitate (tellurous acid, tungstic acid) centrifuged and separated before proceeding with the test.

If thallium(I) is present, the potassium test must be carried out on a portion of the Na_2CO_3 extract which has first been treated with a few drops of H_2S-acetone to remove thallium as thallium(I) sulphide.

Arsenite and arsenate. This test should be omitted because the Na_2CO_3 extract may contain selenite, selenate, tellurite, tellurate and molybdate ions, which also give sulphide precipitates under the conditions of the test.

Selenite and selenate [33]. Dilute 2 or 3 drops of the Na_2CO_3 extract with a few drops of water and add 4M HCl to acidify. Warm to remove CO_2 and cool. Add hydroxylammonium sulphate to saturate the solution, then heat to boiling and allow to stand. *A heavy red precipitate turning dark grey indicates selenite.* When the precipitate has settled, centrifuge and reject the residue. To the solution add a little more hydroxylammonium sulphate and again heat to ensure that all the selenite has been removed. If a further precipitate appears, allow to stand for a few minutes and centrifuge, rejecting the residue. To the clear solution add 0.1 g of $(NH_4)_2Fe(SO_4)_2$ and increase the volume of solution threefold by addition of conc. HCl. Mix and warm, and allow to stand for 1 minute. *A turbid yellowish solution changing to give a brownish red precipitate indicates selenate.*

5.6.3 Cation Analysis

The less common elements are incorporated into the existing Groups (p. 176) as follows:

Silver Group

Tungsten (as white amorphous $H_2WO_4.H_2O$, changing to yellow H_2WO_4 on boiling).
Thallium (as white thallium(I) chloride).

Copper–Tin Group (Note 1)

Selenium (as a yellow mixture of selenium and sulphur, becoming red on heating).
Molybdenum (as brown molybdenum(VI) sulphide) (Note 2).
Tellurium (as a brown mixture of tellurium and sulphur).

Iron Group

Vanadium (as yellow iron(III) vanadate) (see p. 122) [34].
Uranium (as yellow ammonium diuranate).
Zirconium (as white gelatinous hydrated zirconium oxide).
Titanium (as white gelatinous hydrated titanium oxide).
Beryllium (as white gelatinous hydrated beryllium hydroxide).
Thorium (as white hydrated thorium oxide).
Cerium (as yellow hydrated cerium(IV) oxide or white hydrated cerium(III) oxide).

Separation of Cations into Groups
The directions (p. 176) should be amended as follows.

Calcium Group (Note 3)
Add an equal volume of ethanol followed by 10 drops of $2M$ H_2SO_4. Stir well. If a precipitate remains, warm gently for 30 seconds. Stir well. Centrifuge.

Copper-Tin Group
Add 1 drop of $2M$ H_2SO_4 to ensure complete precipitation of the Calcium Group. Centrifuge and reject any residue. Transfer the solution to the beaker and boil off ethanol (Note 4). Cool and add bromine water dropwise until a slight excess is present (yellow solution). Boil off excess of bromine and evaporate until white fumes of SO_3 appear. Cool, dilute with 0.5 ml of water, transfer to a calibrated tube, and wash the beaker with 1 drop of water. Continue as on p. 176.

Iron Group
Evaporate to near dryness. Remove interfering anions except phosphate (Note 9, p. 178). Dilute with 10–15 drops of water. Test for phosphate (Note 10, p. 178), and remove if present (Note 9, p. 178). Add 2 drops of conc. HNO_3 and boil. If the presence of vanadium is suspected (blue colour remaining after Copper-Tin Group), add 4 drops of $1M$ $FeCl_3$. Add 5 drops of 20% NH_4Cl solution, a piece of litmus paper and $4M$ NH_3 solution dropwise till just in excess. Centrifuge.

Zinc Group
(Note 5)

NOTES
1. A blue solution may be formed and sulphur may separate when vanadate is present, because hydrogen sulphide reduces vanadium(V) to vanadium(IV), with formation of sulphur.
2. A transient blue colour is produced before the precipitation of brown molybdenum(VI) sulphide [34].
3. Cerium, zirconium, tellurium and possibly thorium may be partially precipitated in this group. Although sufficient of these elements would probably remain in solution for their detection in the expected groups, by adding a greater amount of $2M$ sulphuric acid than is usual when separating the Calcium Group, it is possible to prevent this precipitation.
4. If a white precipitate is present at this stage, thorium is probably present. The precipitate may be ignored because it will dissolve during the treatment with bromine water.
5. Any thallium(I) not precipitated in the Silver Group precipitates here. It accompanies zinc through the subsequent separation scheme. It should be removed by precipitation with iodide before testing for zinc with hexacyanoferrate(II).

SILVER GROUP

Ag^+, Hg_2^{2+}, Tl^+, $W(VI)$, Pb^{2+}

Wash the precipitate with 0.5 ml of $2M$ HCl (Note). Add 2 ml of water, boil and stir. Centrifuge and separate while hot.

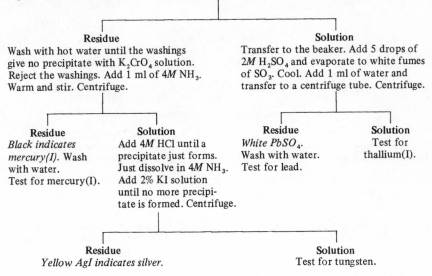

Residue
Wash with hot water until the washings give no precipitate with K_2CrO_4 solution. Reject the washings. Add 1 ml of $4M$ NH_3. Warm and stir. Centrifuge.

Solution
Transfer to the beaker. Add 5 drops of $2M$ H_2SO_4 and evaporate to white fumes of SO_3. Cool. Add 1 ml of water and transfer to a centrifuge tube. Centrifuge.

Residue
Black indicates mercury(I). Wash with water. Test for mercury(I).

Solution
Add $4M$ HCl until a precipitate just forms. Just dissolve in $4M$ NH_3. Add 2% KI solution until no more precipitate is formed. Centrifuge.

Residue
White $PbSO_4$. Wash with water. Test for lead.

Solution
Test for thallium(I).

Residue
Yellow AgI indicates silver.

Solution
Test for tungsten.

Confirmatory Tests

Mercury(I)
    Test as on p. 179.

Lead
    Dissolve in $2M$ ammonium acetate. Add 1 drop of 5% K_2CrO_4 solution. *A yellow precipitate indicates lead.*

Thallium(I)
    Add 2 or 3 drops of 10% $Na_2S_2O_3$ solution followed by 1 or 2 drops of 2% KI solution. *A bright yellow precipitate indicates thallium.*

Tungsten(VI)
    Evaporate to about 0.5 ml. Add 3 drops of 5% $SnCl_2$ solution and 3 drops of conc. HCl. Heat to boiling. *A blue colour or blue precipitate indicates tungsten.*

NOTE
If the precipitate is washed with water, any tungstic acid present may form a colloidal solution. A clear supernatant solution is not then obtained on centrifuging. See also Note 5 on p. 177.

The separation of the Calcium Group and the parting of the Copper–Tin Group remain the same as already described (Calcium Group p. 180, parting pp. 182 and 186). The tests given here refer to the hydroxide parting; the chloroacetate procedure is given on p. 208.

COPPER GROUP

Cu^{2+}, Bi^{3+}, Cd^{2+}, Hg^{2+}, Se(IV), Se(VI)

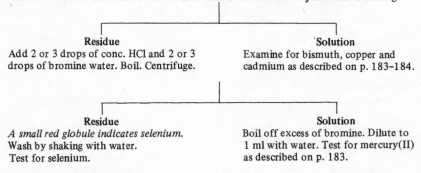

Wash the residue *twice* with 1 ml of water. Add 1 ml of $4M$ HNO_3 and boil. Centrifuge.

Residue	**Solution**
Add 2 or 3 drops of conc. HCl and 2 or 3 drops of bromine water. Boil. Centrifuge.	Examine for bismuth, copper and cadmium as described on p. 183–184.

Residue	**Solution**
A small red globule indicates selenium. Wash by shaking with water. Test for selenium.	Boil off excess of bromine. Dilute to 1 ml with water. Test for mercury(II) as described on p. 183.

Confirmatory Test

Selenium

The presence and oxidation state of selenium should already have been established (p. 203). However, the presence of selenium may be confirmed as follows. Dry the residue by shaking it twice with acetone. Remove the residue and press it lightly between filter papers. Add the residue to a dry centrifuge tube in which a few drops of conc. H_2SO_4 have just been heated to boiling. Boil for 30 seconds. Allow to cool. A *green colour,* changing to a *red colour or red precipitate on dilution with water* (**add with care!**)*, indicates selenium.*

TIN GROUP

Sn(II), Sn(IV), Sb(III), As(III), Te(IV), Te(VI), Mo(VI)

Just acidify the KOH extract with $2M$ HCl and add 4–6 drops of H_2S–acetone. Centrifuge and discard the solution. Wash the residue. Add 2 ml of $4M$ HCl. Boil for 5–10 seconds. Cool. Add a few drops of H_2S–acetone. Centrifuge.

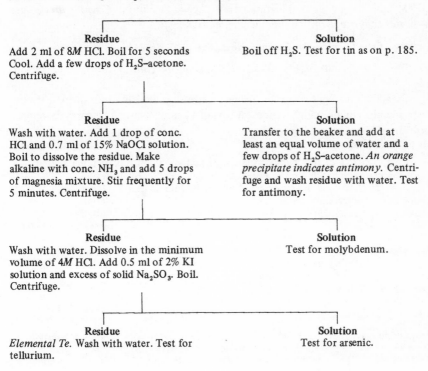

Residue
Add 2 ml of $8M$ HCl. Boil for 5 seconds
Cool. Add a few drops of H_2S–acetone.
Centrifuge.

Solution
Boil off H_2S. Test for tin as on p. 185.

Residue
Wash with water. Add 1 drop of conc.
HCl and 0.7 ml of 15% NaOCl solution.
Boil to dissolve the residue. Make
alkaline with conc. NH_3 and add 5 drops
of magnesia mixture. Stir frequently for
5 minutes. Centrifuge.

Solution
Transfer to the beaker and add at
least an equal volume of water and a
few drops of H_2S–acetone. *An orange
precipitate indicates antimony.* Centrifuge and wash residue with water. Test
for antimony.

Residue
Wash with water. Dissolve in the minimum
volume of $4M$ HCl. Add 0.5 ml of 2% KI
solution and excess of solid Na_2SO_3. Boil.
Centrifuge.

Solution
Test for molybdenum.

Residue
Elemental Te. Wash with water. Test for
tellurium.

Solution
Test for arsenic.

Confirmatory Tests

Molybdenum

Acidify with conc. HCl. Add 2 drops of $2M$ NH_4SCN, a few drops of ether and 1 drop of 5% $SnCl_2$ solution. Stir well. *A red colour in the ether layer indicates molybdenum.*

Arsenic

Boil with conc. HNO_3. Add excess of molybdate reagent and boil. *A yellow precipitate indicates arsenic.*

Tellurium

Add a few drops of conc. H_2SO_4 and heat. Cool. *A red colour,* changing to a *grey colour or precipitate on dilution with water, indicates tellurium.*

The oxidation state of tellurium in the original material should be determined by *carefully* neutralizing 3 or 4 drops of the Na_2CO_3 extract with $4M$ HNO_3 and adding 2 or 3 drops of 1% $AgNO_3$ solution. *A white precipitate indicates tellurite; a brown precipitate, tellurate.*

Parting of the Copper–Tin Group (alternative procedure).

Test for Tin Group Metals Dissolved by Chloroacetate Solution (As, Sb, Sn, Mo, Te).

In addition to the tests for arsenic, antimony and tin given on p. 186, the tests for molybdenum and tellurium described above should be carried out on a portion of the solution obtained.

IRON GROUP

Fe(III), Al(III), Cr(III), Zr(IV), Ti(IV), Ce(III), Th(IV), U(VI), V(IV), V(V), Be(II)

Wash the precipitate with 1 ml of water. Centrifuge and discard the washings. Add 2 or 3 drops of $4M$ NaOH and 10–12 drops of water. Boil for at least 1 minute (Note 1). Cool (Note 2). Centrifuge.

Residue	**Solution**
Dissolve in $2M$ H_2SO_4. Heat to boiling and add 1 drop of 10% Na_2HPO_4 solution. If a precipitate is formed, continue addition until precipitation is complete (Note 3). Centrifuge.	Add a small piece of litmus paper followed by $4M$ HCl dropwise until neutral. Centrifuge.

Residue	**Solution**	**Residue**	**Solution**
Wash with a few drops of $2M$ H_2SO_4. Centrifuge and discard washings. Add 2 or 3 drops of $2M$ H_2SO_4 and 5–10 drops of 20 vol. H_2O_2. Stir well. Centrifuge (Note 3).	Add zirconium nitrate solution dropwise until precipitation is complete. Centrifuge, and discard the precipitate (Note 4). Add a piece of litmus paper followed by $4M$ NH_3 until alkaline. Centrifuge and discard the solution. Wash the precipitate with a few drops of water. Centrifuge, and discard the washings. To the residue add 4 drops of $4M$ NaOH and 4 drops of 20-vol. H_2O_2. Boil for 30 seconds. Centrifuge.	*Al(OH)$_3$*. Wash with water. Test for aluminium.	Test for vanadium.

Residue	**Solution**
Zirconium or titanium phosphate Wash with $2M$ H_2SO_4. Test for zirconium.	*Yellow indicates titanium.*

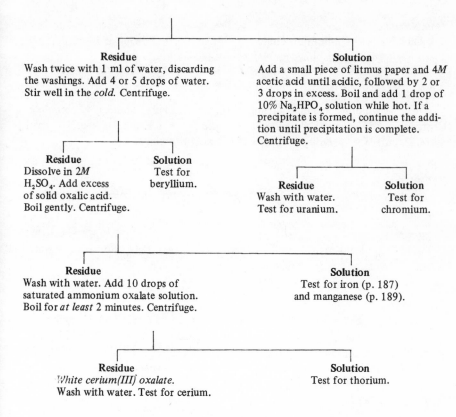

Residue
Wash twice with 1 ml of water, discarding the washings. Add 4 or 5 drops of water. Stir well in the *cold*. Centrifuge.

Solution
Add a small piece of litmus paper and 4*M* acetic acid until acidic, followed by 2 or 3 drops in excess. Boil and add 1 drop of 10% Na_2HPO_4 solution while hot. If a precipitate is formed, continue the addition until precipitation is complete. Centrifuge.

Residue
Dissolve in 2*M* H_2SO_4. Add excess of solid oxalic acid. Boil gently. Centrifuge.

Solution
Test for beryllium.

Residue
Wash with water. Test for uranium.

Solution
Test for chromium.

Residue
Wash with water. Add 10 drops of saturated ammonium oxalate solution. Boil for *at least* 2 minutes. Centrifuge.

Solution
Test for iron (p. 187) and manganese (p. 189).

Residue
White cerium(III) oxalate. Wash with water. Test for cerium.

Solution
Test for thorium.

Confirmatory Tests

Aluminium
    Dissolve in the minimum volume of 4*M* NaOH solution. Test as on p. 187.

Vanadium
    To 4 drops of the solution add 4 drops of conc. HCl. Add an equal volume of acetylacetone solution, shake, and allow to stand for one minute. *A permanent blue top layer indicates vanadium.*

Beryllium
    Add 3 or 4 drops of 0.025% *p*-nitrobenzeneazo-orcinol solution. *A red colour indicates beryllium.*

Uranium

Dissolve in the minimum volume of 4M HCl. Add 2 drops of 5% $K_4Fe(CN)_6$ solution. *A brown precipitate indicates uranium.*

Chromium

Add zirconium nitrate solution dropwise until precipitation is complete. Centrifuge and discard the precipitate. To the solution add 2 drops of 10% lead acetate solution. *A yellow precipitate indicates chromium.*

Thorium

Add a few drops of 4M HCl. *A white crystalline precipitate indicates thorium.*

Cerium

Add 0.5 ml of 4M NH$_3$ and 2 or 3 drops of 20 vol. H_2O_2. Boil for 1 minute. *A yellow precipitate indicates cerium.*

Zirconium

Centrifuge and discard the washings. Add 4 drops of saturated ammonium oxalate solution and a few small crystals of oxalic acid. Add 4M NaOH until precipitation is complete. Centrifuge, wash with three 1-ml portions of 0.04M NaOH. Centrifuge and reject the washings. Dissolve the residue in 3 drops of conc. HCl and add 5 drops of 10% mandelic acid solution. *A white precipitate indicates zirconium.*

NOTES
1. Beryllium hydroxide initially dissolves as tetrahydroxoberyllate ions, then decomposes hydrolytically. Provided the solution is sufficiently dilute, the beryllium is quantitatively reprecipitated as the hydroxide.
2. In hot solution, hydrated titanium(IV) oxide tends to give a colloidal suspension, especially if only a small amount is present. It is more readily centrifuged on cooling.
3. If phosphate was removed with zirconium nitrate before the separation of the Iron Group, then zirconium will be found here. Its presence in the original sample must be ascertained by carrying out a similar analysis without adding zirconium nitrate.
4. The added phosphate must be removed at this stage. If not removed, Iron Group metals remaining will precipitate as phosphates when the solution is made alkaline.

5.6.4 Anion Analysis

Many of the less common elements may be found in the sodium carbonate extract as anions. These include selenite (SeO_3^{2-}), selenate (SeO_4^{2-}), tellurite (TeO_3^{2-}) and tellurate ($H_4TeO_6^{2-}$) as well as hydroxo- or oxy-anions of metals [MoO_4^{2-}, WO_4^{2-}, VO_3^-, $U_2O_7^{2-}$ and $Be(OH)_4^{2-}$]. These are accommodated in the separation scheme for anions as described below.

SEPARATION SCHEME FOR ANIONS

Add 1 ml of the sodium carbonate extract to the silver ammine solution as described on p. 192 (Notes 1 and 2). Centrifuge.

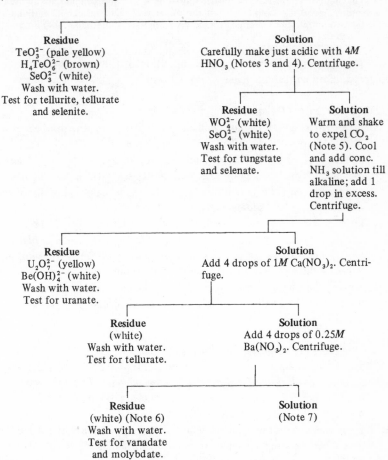

Residue
TeO_3^{2-} (pale yellow)
$H_4TeO_6^{2-}$ (brown)
SeO_3^{2-} (white)
Wash with water.
Test for tellurite, tellurate and selenite.

Solution
Carefully make just acidic with $4M$ HNO_3 (Notes 3 and 4). Centrifuge.

Residue
WO_4^{2-} (white)
SeO_4^{2-} (white)
Wash with water.
Test for tungstate and selenate.

Solution
Warm and shake to expel CO_2 (Note 5). Cool and add conc. NH_3 solution till alkaline; add 1 drop in excess. Centrifuge.

Residue
$U_2O_7^{2-}$ (yellow)
$Be(OH)_4^{2-}$ (white)
Wash with water.
Test for uranate.

Solution
Add 4 drops of $1M$ $Ca(NO_3)_2$. Centrifuge.

Residue
(white)
Wash with water.
Test for tellurate.

Solution
Add 4 drops of $0.25M$ $Ba(NO_3)_2$. Centrifuge.

Residue
(white) (Note 6)
Wash with water.
Test for vanadate and molybdate.

Solution
(Note 7)

NOTES

1. Best results are obtained if the sodium carbonate extract is carefully neutralized with dilute nitric acid before it is added to the silver ammine solution. During the pH adjustment many less common anions give transient precipitates which redissolve when the correct pH is attained.
2. Vanadium(IV) passes into the sodium carbonate extract and reduces the silver ammine to silver metal [34]. This can be avoided if all the vanadium is oxidized to vanadium(V) with hydrogen peroxide by the procedure described for the destruction of cyanide, thiosulphate and permanganate (p. 191).
3. None of the less common species is extracted into benzene.
4. Because excess of nitric acid dissolves silver selenate, the solution is only slightly acidified. Thus, if oxalate is present, it may precipitate with the selenate. However, it does not interfere with the test for selenate.

5. If vanadate is present it gives a slight brown precipitate on warming and later on addition of calcium nitrate; the yellow supernatant liquid is very dark [34].
6. Warming produces more vanadate precipitate [34].
7. Thallium(I) carbonate is somewhat soluble in water, so thallium may be found in the sodium carbonate extract, and in the solution at this stage. Thallium(I) interferes in the test for citrate by producing a brown precipitate of manganese(IV) oxide.

Tests on the Precipitates

The tests given below for the examination of the precipitates must be used in conjunction with those given on pp. 193–197.

Tests for Tellurite, Tellurate and Selenite [35]

Add 1 ml of 2*M* HCl, shake well and centrifuge. Test the solution as follows.
A. Add 3 drops of conc. HCl to 6 drops of the solution, followed by a small quantity of solid Na₂SO₃. *An immediate, black precipitate indicates tellurite.* If no precipitate appears within 10 seconds, warm slightly. *A red precipitate indicates selenite.* If no precipitate is formed, boil for 30 seconds. *A black precipitate indicates tellurate* (Note 1).
B. To 6 drops of the solution add 10% NH₄I solution dropwise. *A red precipitate accompanied by liberation of iodine indicates selenite. A black precipitate, soluble in an excess of iodide to give a red-brown solution, indicates tellurite.* Add an excess of 10% Na₂S₂O₃ solution to decolourize the red-brown solution. *A red precipitate indicates selenite. Tellurate* gives no reaction in the cold but on warming is reduced and gives the same reaction as tellurite.

Tests for Tungstate and Selenate (Bromate, Iodate) [35]

A. *Selenate.* To a portion of the residue add 0.5 ml of 8*M* HCl and boil for 30 seconds. *Evolution of chlorine indicates selenate, bromate and/or iodate.* Cool, remove the AgCl precipitate and add solid Na₂SO₃ to the solution. *A red precipitate of selenium confirms selenate.*
B. Treat the rest of the precipitate as for bromate and iodate (p. 195), except that when decomposing the ammonium sulphide, add 5 or 6 drops of 4*M* HClO₄. Continue heating until H₂S is completely removed. Centrifuge.

Residue	**Solution**
Wash with water. Add 3 drops of 5% SnCl₂ solution and 5 drops of conc. HCl. Heat to boiling. *A blue precipitate indicates tungstate.*	Test for bromide and iodide as described on p. 194.

Test for Uranate

Dissolve the precipitate in the minimum volume of 4*M* acetic acid. Add 2 drops of 5% K₄Fe(CN)₆ solution. *A brown precipitate, changing to yellow on treatment with excess of 4M NaOH, indicates uranate.*

Test for Tellurate on the Tellurate Precipitate [35]

Dissolve the precipitate in 5% acetic acid. Take 6 drops, add 3 drops of conc. HCl and a small amount of solid Na_2SO_3. Boil for 30 seconds. *A black precipitate of tellurium indicates tellurate* (Note 2).

Test for Vanadate and Molybdate (Chromate, Sulphate)

Add 0.5 ml of $4M$ HCl, warm and centrifuge. *A white precipitate remaining indicates sulphate.* Centrifuge. Dilute the solution to 1 ml. Divide into 3 parts and test as follows.

Chromate

As on p. 196.

Molybdate

Add 1 or 2 drops of $2M$ NH_4SCN, a few drops of ether and 1 drop of 5% $SnCl_2$ solution. Stir well. *A red ether layer indicates molybdate.*

Vanadate

As on p. 209 (Note 3).

NOTES
1. If selenite or tellurite is present, tellurate is more conveniently confirmed by using the calcium tellurate precipitate.
2. The sensitivity of the test is reduced if phosphate is also present.
3. Chromium gives a transient violet colour in the upper layer, but a green lower layer.

5.7 ANALYSIS OF INSOLUBLE SUBSTANCES [36]

The insoluble substances dealt with here are those which do not dissolve in the common aqueous acids. Those most likely to be encountered are the following:

Silver salts	AgCl, AgBr, AgI, AgSCN
Sulphates	$PbSO_4$, $BaSO_4$, $SrSO_4$
Ignited oxides	Fe_2O_3, Al_2O_3, Cr_2O_3, SnO_2
Other oxides	WO_3, MoO_3, TiO_2, ZrO_2, SiO_2
Others	CaF_2 and other insoluble fluorides, $Cu_2Fe(CN)_6$, fused $PbCrO_4$, W, C, Si, SiC, BN, silicate minerals, $Zr_3(PO_4)_4$, $Ti_3(PO_4)_4$.

The scheme involves a number of preliminary tests on the original sample, and the isolation and a systematic analysis of the insoluble material.

ISOLATION OF THE INSOLUBLE MATERIAL

Treat 50 mg of the sample with 2 ml of $4M$ HNO_3 and boil for 2 minutes.

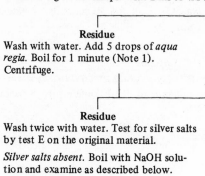

Residue
Wash with water. Add 5 drops of *aqua regia*. Boil for 1 minute (Note 1). Centrifuge.

Solution
Add the solution obtained below. Evaporate half the solution to dryness. Analyse any residue by the usual cation scheme

Residue
Wash twice with water. Test for silver salts by test E on the original material.

Silver salts absent. Boil with NaOH solution and examine as described below.

Silver salts present. Add 1 ml of freshly filtered 50% $(NH_4)_2S_2O_3$ solution and stand with occasional stirring. Centrifuge.

Solution
Add to the solution already obtained.

Residue
Wash with water. Boil with NaOH solution and examine as described below.

Solution
Add $4M$ HNO_3 and identify the precipitated silver salts as on pp. 193–194.

Preliminary Tests

If insoluble materials have been isolated (above) test the *original material* as follows (Note 2), unless the test has already been carried out.

A. *Colour.* W,Si,C (black); SiC (grey-black); Cr_2O_3 (green); WO_3, AgBr, AgI (yellow); Fe_2O_3, $Cu_2Fe(CN)_6$ (dark brown); $PbCrO_4$ (orange-red). Some silicate minerals may be coloured by trace impurities.

B. *Flame Test.* Heat on the platinum wire in a reducing flame. Moisten with conc. HCl and reheat. *Calcium, barium, strontium* and *copper* may be detected p. 165).

C. *Microcosmic Salt Bead.* (p. 166). *Silica* and *silicates* give a typical 'skeleton'. *Chromium* gives a green bead, *titanium* violet, and *tungsten* pale blue.

D. *Charcoal Block Reaction* (p. 165). If *tin* is detected, transfer some of the fused material to a beaker, add 1 ml of conc. HCl and stir. Add a small piece of zinc. Immerse in the solution the round end of a test-tube containing water, then hold this part of the tube in a non-luminous flame. *A blue colour in the flame near the tube surface confirms the presence of tin.*[†]

E. *Test for Silver Salts.* Shake 10 mg of sample with 0.5 ml of $K_2Ni(CN)_4$ solution and 2 drops of 1% dimethylglyoxime solution. *A red precipitate indicates the presence of silver salts.*

F. *Test for Insoluble Fluorides.* Add 1 drop of $Zr(NO_3)_4$-alizarin reagent to 5 mg of the sample. *A change from red-violet to yellow indicates an insoluble fluoride.*

†The Meissner test (H. Meissner, *Z. Anal. Chem.*, 1930, **80**, 247).

G. *Test for Boron Nitride.* Test for *boron* as on p. 171. Place 20-30 mg of the sample in a beaker with 2 drops of conc. H_2SO_4. Heat to fumes. Cool, make alkaline with $4M$ NaOH and boil. *The evolution of ammonia indicates nitride.*

BOILING WITH SODIUM HYDROXIDE SOLUTION

Add 1 ml of $4M$ NaOH and boil. *The rapid evolution of hydrogen indicates silicon.* Boil until reaction is complete and the black colour has gone (Note 3). Centrifuge.

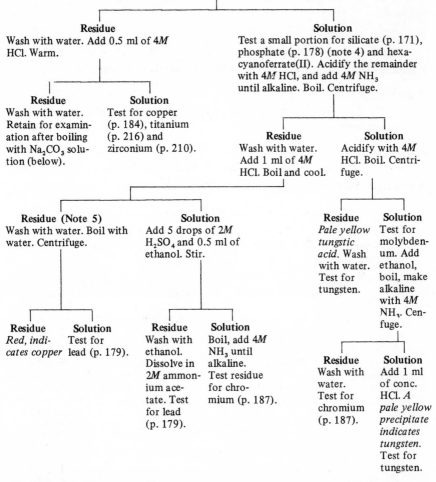

Residue
Wash with water. Add 0.5 ml of $4M$ HCl. Warm.

Residue
Wash with water. Retain for examination after boiling with Na_2CO_3 solution (below).

Solution
Test for copper (p. 184), titanium (p. 216) and zirconium (p. 210).

Solution
Test a small portion for silicate (p. 171), phosphate (p. 178) (note 4) and hexacyanoferrate(II). Acidify the remainder with $4M$ HCl, and add $4M$ NH_3 until alkaline. Boil. Centrifuge.

Residue
Wash with water. Add 1 ml of $4M$ HCl. Boil and cool.

Solution
Acidify with $4M$ HCl. Boil. Centrifuge.

Residue (Note 5)
Wash with water. Boil with water. Centrifuge.

Solution
Add 5 drops of $2M$ H_2SO_4 and 0.5 ml of ethanol. Stir.

Residue
Red, indicates copper

Solution
Test for lead (p. 179).

Residue
Wash with ethanol. Dissolve in $2M$ ammonium acetate. Test for lead (p. 179).

Solution
Boil, add $4M$ NH_3 until alkaline. Test residue for chromium (p. 187).

Residue
Pale yellow tungstic acid. Wash with water. Test for tungsten.

Solution
Test for molybdenum. Add ethanol, boil, make alkaline with $4M$ NH_3. Cenfuge.

Residue
Wash with water. Test for chromium (p. 187).

Solution
Add 1 ml of conc. HCl. *A pale yellow precipitate indicates tungsten.* Test for tungsten.

Confirmatory Tests
Silicate

Place 2 drops of $(NH_4)_6Mo_7O_{24}$–HNO_3 solution in a crucible. Warm until white fumes appear. Add one drop of the test solution and cool. Add 1 drop of

o-tolidine solution and 2 drops of conc. NH_3 solution. *A blue colour indicates silicate.*

Hexacyanoferrate(II)

Acidify 2 drops of test solution with $4M$ HCl. Add one drop of 2% $FeCl_3$ solution. *A dark blue precipitate indicates hexacyanoferrate(II).*

Molybdenum

Add to 3 drops of the test solution, 1 drop of 2% 2,2'-bipyridyl solution and 1 drop of 10% $SnCl_2$ solution in $4M$ HCl. *A violet precipitate indicates molybdenum.*

Tungsten

Dissolve residue in $4M$ NH_3. Add 2 drops of 10% $SnCl_2$ solution in $4M$ HCl. Warm. Allow to stand for a minute. *A blue colour indicates tungsten.*

Titanium

Add 1 drop of 20 vol. H_2O_2. *A yellow colour that disappears on adding NaF indicates titanium.*

Zirconium

Add 5 drops of 20 vol. H_2O_2 and 3 drops of 10% Na_2HPO_4 solution. Boil. *A white gelatinous precipitate indicates zirconium.*

BOILING WITH SODIUM CARBONATE SOLUTION

Transfer the residue from boiling with NaOH solution (above) to a beaker, with 1–2 ml of water. Add 5 times its bulk of anhydrous Na_2CO_3. Boil for 5 minutes. Cool. Centrifuge.

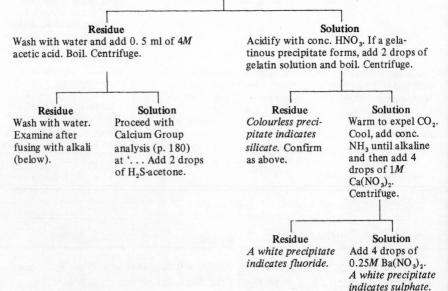

FUSIONS WITH ALKALI AND PYROSULPHATE

Dry the residue from boiling with Na_2CO_3 solution. Transfer it to a nickel spoon and mix it well with 3 times its bulk of 1:1 Na_2CO_3: $NaNO_3$ mixture (Note 6). Heat strongly for 1 minute, *allow to cool* and place in a beaker containing 2 ml of water. Boil for 1 minute. Centrifuge.

Residue	Solution
Wash with water. Dry, mix with 30 mg of powdered $KHSO_4$ or $K_2S_2O_7$ and fuse for 30 seconds in a crucible. Cool. Add 1 ml of $2M$ H_2SO_4, stir and decant into a centrifuge tube. Centrifuge. Test the solution for zirconium and titanium (as above).	Test small portions for silicate and tungstate (as above) and for borate (below) (Note 7). Acidify the remainder with conc. HCl, add 0.5 ml in excess, and a few drops of gelatin solution. Evaporate to dryness, add 0.5 ml of conc. HCl and again evaporate to dryness. Add 1 ml of water and 2 drops of $4M$ HCl. Warm. Reject any residue. Add H_2S–acetone.

Residue	Solution
Test for tin (p. 185).	Boil off H_2S, remove borate (p. 178) and add $4M$ NH_3 until alkaline. Analyse the precipitate by the Iron Group scheme (p. 208) (Note 8).

Test for Borate. Evaporate 5 drops to dryness in a crucible. Add 5 drops of 0.1% alizarin solution in conc. H_2SO_4, then add a drop of water. *The colour changes from yellow to red if borate is present.*

NOTES
1. If pyrolusite (MnO_2) is present, boil until all has dissolved, if necessary adding more *aqua regia*.
2. The use of a large number of preliminary tests is justified because the nature of the insoluble material can often rapidly be deduced before it is confirmed by the later systematic analysis.
3. If the solution is yellow, lead chromate is present. The solution must then be boiled for one minute. Most of the lead chromate dissolves, but a little green chromium(III) oxide remains.
4. The presence of phosphate indicates the presence of zirconium or titanium phosphate in the original material.
5. If silicon is present, the residue will contain hydrated silica. Proceed as follows. Add 1 ml of $4M$ HCl and evaporate to dryness. Repeat. Boil again with 1 ml of $4M$ HCl.

Residue	Solution
Wash with water. Test for silicate.	Add 5 drops of $2M$ H_2SO_4 and 1 ml of ethanol. Centrifuge.

Residue	Solution
Wash with ethanol. Test for lead (p. 179).	Boil off the ethanol, add $4M$ ammonia until alkaline. Test residue for chromium (p. 187).

6. This fusion mixture is adequate for all the compounds remaining except alumina, titania and zirconia. If difficulty is experienced in the fusion of alumina, fuse with *fresh* sodium peroxide. Note, however, that this reagent has been known to give rise to explosions, and that there is considerable attack on the nickel spoon.
7. If silicate and tungstate are absent, merely acidify with 4*M* hydrochloric acid and add H_2S-acetone.
8. On boiling a hydrated chromium(III) oxide precipitate with sodium hydroxide solution and hydrogen peroxide, any white residue will be hydrated silica or tungstic acid not removed previously.

REFERENCES

The references listed below are to the original MAQA investigations, all of which are published in *Mikrochimica Acta*.

[1] Belcher, R., **1956**, 1842, (Part I).

[2] Belcher, R., Harrison, R. and Stephen, W. I., **1958**, 201 (Part VI).

[3] Hayes, O. B. and Winterburn, J., **1958**, 197, (Part V).

[4] Hayes, O. B., **1960**, 366, (Part XV).

[5] Evans, D. L., **1967**, 386, (Part XXVI).

[6] Rockett, B. W. and Kirby, M., **1968**, 1277, (Part XXXV).

[7] Falkner, P. R. and Burns, D. T., **1965**, 318, (Part XXI).

[8] Falkner, P. R. and Burns, D. T., **1965**, 322, (Part XXII).

[9] Catchpole, A. G., Kirby, M., Dowson, W. M. and Williams, M., **1968**, 1269, (Part XXXIII).

[10] Burns, D. T. and Drake, G. H., **1967**, 390, (Part XXVII).

[11] Jones, W. F., **1961**, 214, (Part XVIII).

[12] Stephen, W. I., **1960**, 928, (Part XVI).

[13] Dowson, W. M., **1959**, 841, (Part XIII).

[14] Bark, L. S., **1958**, 117, (Part III).

[15] Jones, W. F., **1959**, 635, (Part XII).

[16] Beardsley, D. A., Briscoe, G. B., Clark, E. R., Matthews, A. G. and Williams, M., **1970**, 1287, (Part XLI).

[17] Bailey, D., Dowson, W. M., Harrison, R. and West, T. S., **1958**, 137, (Part IV).

[18] Jones, W. F., **1959**, 544, (Part IX).

[19] Bailey, D. and Dowson, W. M., **1960**, 12, (Part XIV).

[20] Bark, L. S. and Jones, W. F., **1958**, 406, (Part VII).

[21] Stephen, W. I. and Weston, A. M., **1964**, 179, (Part XIX).

[22] Belcher, R. and Weisz, H., **1956**, 1847, (Part II).

[23] Belcher, R. and Weisz, H., **1958**, 571, (Part VIII).

[24] Belcher, R., and Stephen, W. I., **1959**, 547, (Part X).

[25] Catchpole, A. G., **1968**, 586, (Part XXXI).

[26] Hadlington, M. and Rockett, B. W., **1967**, 238, (Part XXV).

[27] Hayes, O. B., **1968**, 647, (Part XXXII).

[28] Osborne, V. J. and Freke, A. M., **1964**, 790, (Part XX).

[29] Dowson, W. M. and Jones, W. F., **1974**, 339, (Part XLIV).

[30] Jones, W. F., **1961**, 88, (Part XVII).

[31] Burns, D. T. and Lee, J. D., **1969**, 206, (Part XXXVII).

[32] Burns, D. T., Lee, J. D. and Harris, L. G., **1972**, 188, (Part XLII).

[33] Dowson, W. M., **1969**, 202, (Part XXXVI).

[34] Carter, A. H., **1970**, 20, (Part XXXIX).

[35] Falkner, P. R., **1968**, 1223, (Part XXXIV).

[36] Jones, W. F., **1967**, 1019, (Part XXX).

Chapter 6

Ring-Oven Technique in Qualitative Inorganic Analysis

The ring-oven technique provides an elegant and rapid method of analysis which has the additional advantage of using only small quantities of samples and reagents. It is a particular kind of spot analysis carried out on filter paper on which the substances to be detected or determined are concentrated in the form of well-defined circular lines. The ring-oven was developed originally for the separations of extremely minute samples, and has since been applied to inorganic and organic qualitative and quantitative analysis, and has also been combined with other analytical techniques, such as radiochemical methods [1]. Because it enables analyses to be carried out on single drops of solution a brief account of the uses of the ring-oven in qualitative analysis is included in this book.

It is hardly possible to achieve the qualitative analysis of a single drop (about 50 μl) of sample solution by the conventional methods. However, it is possible to place the drop on a filter paper and to fix one substance (or group of substances) of the sample as a precipitate within the pores of the filter paper by addition of a suitable reagent. The soluble components of the sample drop can then be washed to an outer zone of the paper. However, with the small volume of solvent used, 50–100 μl, separation is unlikely to be complete. Also, if a larger volume of solvent is used to wash all the soluble components completely from the precipitate, then the soluble components are distributed, much diluted, over a large irregular area of the filter paper.

In order to make this technique more generally applicable, the ring-oven method has been developed, by which soluble materials can be completely removed from a precipitate fixed on filter paper, and concentrated in a sharply defined ring zone.

6.1 THE RING-OVEN AND ITS OPERATION

Figure 6.1 shows a schematic diagram of a ring-oven: it is basically a cylindrical aluminium block H (55 mm in diameter, 35 mm high) with a vertical hole 22 mm in diameter (dotted line) bored through its centre. A suitably insulated heating wire is installed in the block and a variable transformer or resistor is used to regulate the temperature. The temperature of the top surface of the block should be 105–110° for work with aqueous solutions (or in general about 10° above the boiling point of the solvent used for washing out). The glass tube Gl (50 mm long) serves as a guide tube for a capillary pipette which just fits into it. The tip of the pipette must be carefully rounded; this can easily be done by grinding the capillary, water-filled, in a porcelain mortar. The capillary pipette must be vertical with its centre coincident with that of the hole in the block and must end a few mm above the surface of the block. Adjustment is by means of the three screws, S. The pipette is centred by allowing it to rest on a small metal block which fits into the hole of the ring-oven and has a central fiduciary mask (Fig. 6.2). The small electric bulb below the heating block illuminates the paper and allows the washing out procedure to be more readily seen and controlled.

The following description and example illustrate the mode of operation of the ring-oven.

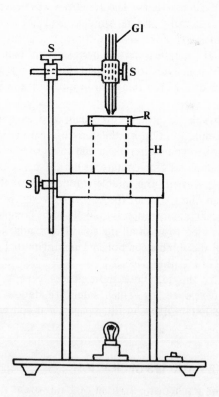

Fig. 6.1 – Schematic diagram of a ring-oven.

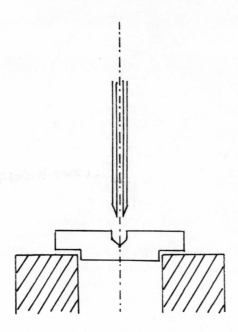

Fig. 6.2 – Centring of pipette.

Place a drop of an iron(III) chloride solution (100 ppm) in the centre of a 55-mm diameter filter paper (Whatman No. 40, Schleicher & Schüll 589 [2] or Macherey Nagel MN 2260 are most commonly used) and place the filter paper centrally on the hot ring-oven. Keep the filter paper in position by means of a porcelain, glass or metal ring, R (Fig. 6.1), inner diameter about 25 mm. Using 0.05M hydrochloric acid in a capillary pipette (about 0.1 mm bore), wash out all the iron(III) chloride. Fill the pipette by touching it on the surface of the hydrochloric acid, then place it carefully through the glass tube Gl onto the iron(III) chloride spot. The filter paper absorbs the solvent and the wet spot spreads concentrically. Refill the pipette and again place it on the spot.

As soon as the liquid approaches the edge of the hole in the block the solvent vaporizes. The diameter of the moist spot cannot exceed that of the hole (22 mm) and the iron(III) chloride collects as a sharply defined ring zone provided the surface temperature of the block is correct. Repeat the washing about eight times. All the iron(III) chloride will now be concentrated in the ring zone. The whole procedure takes 1–2 minutes. Dry the filter paper with a hot-air drier (or in a drying oven) and spray it with a 1% aqueous solution of potassium hexacyanoferrate(II). A well-defined ring of Prussian Blue appears where the iron has been concentrated. The inner area of the paper is completely free from iron and therefore is not coloured by the reagent spray (see Fig. 6.3).

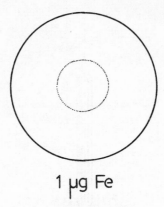

1 μg Fe

Fig. 6.3 – Ring of Prussian Blue.

The ring zone is as thin as a fine pencil line. If the width of the ring zone were as great as 1 mm, the area of the ring would be less than 70 mm² and therefore smaller than that of a spot of 10 mm diameter (≡1.5 μl sample volume). However, the ring zone is normally even narrower (0.1–0.3 mm), and the area of the ring zone is thus only about 7–20 mm². Hence the concentration of the ions in the ring is 3–10 times that in the original spot. Most of the tests carried out on the ring are therefore more sensitive than simple paper spot-tests.

If several ions are to be detected in a single sample, without separation, the drop of sample solution is eluted into the ring zone, which can then be divided into as many as 10 sectors. Each sector, after treatment with a suitable reagent, will show a circular arc at least 7 mm long. As the subdivision of the sample has been accomplished without dilution, a high sensitivity is achieved for the identification reactions.

6.2 SEPARATIONS

When a separation is to be carried out, one group of ions is precipitated in the centre of the filter paper and the unprecipitated ions are washed into a ring zone as described above. For this precipitation gaseous reagents are used, whenever possible, to avoid the danger of enlarging the initial spot.

A gas generator (Fig. 6.4) has been designed for this purpose. The 50-ml glass flask is connected to a dropping funnel, D, and two glass tubes, O and U, with flat-ground flanges. The flanges are held together by means of spiral springs, S. The apparatus is connected to a water-pump by the stopcock, C_2. The procedure for the analysis of a mixture containing copper, iron and nickel may be used to illustrate the separation of ions into two groups.

Mark the middle of the filter paper with a pin. Transfer an aliquot of the solution onto the marked centre by means of a self-filling capillary sample pipette (Fig. 6.5). These pipettes are used so that drops of constant size (about

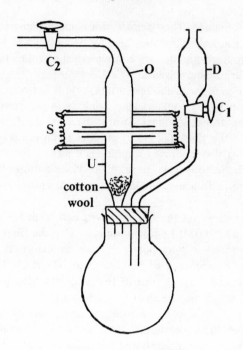

Fig. 6.4 – Gas generator.

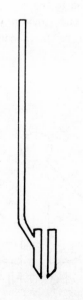

Fig. 6.5 – Self-filling pipette.

1.5 μl) are always obtained. They consist of a conically tapered capillary tube attached to a glass rod holder and are filled by being touched to the surface of the sample solution. Such pipettes are easily cleaned by filling them a few times with alcohol or other suitable solvent.

Place the filter paper with the test solution on it between the two flanges of the gas generator, which contains zinc sulphide in the glass flask and 12M sulphuric acid in the dropping funnel. Turn stopcock C_1 to allow a few drops of the acid to fall onto the zinc sulphide to produce hydrogen sulphide. Carefully open stopcock C_2 so that hydrogen sulphide is drawn through the filter paper and precipitates all the copper as its sulphide. Place a drop of alcohol on the spot and repeat the hydrogen sulphide treatment to ensure complete precipitation.

Place the filter paper on the hot ring-oven and wash the iron and nickel into the ring zone with 0.05M hydrochloric acid. Dry the filter and cut out the inner spot with a stainless steel punch 12 mm in diameter. This operation should be done on a paper-covered wooden base. The small disc cut out contains all the copper, and the ring zone of the remaining filter paper contains all the iron and nickel, which may be identified as follows.

Cut out two sectors of the ring zone and spray one of them with a 1% solution of potassium hexacyanoferrate(II); a blue line appears if iron is present. Fume the second sector over concentrated ammonia and spray with a 1% solution of dimethylglyoxime in ethanol: a red line indicates nickel. Moisten the inner disc with a drop of alcohol and hold it in a wide-necked bottle containing bromine water, in order to oxidize the sulphide ions to sulphate. Then fume the filter disc over ammonia and spray it with a 1% ethanolic solution of rubeanic acid: an olive-green to black spot indicates copper (Fig. 6.6).

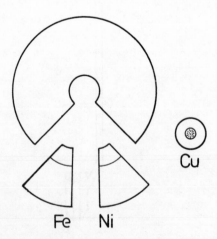

Fig. 6.6 – Detection of iron, nickel and copper.

The basic procedure also makes it possible to separate ions into more than two groups. An example is the separation of lead, copper and aluminium.

Place the sample solution on the filter paper, and treat it with hydrogen sulphide as in the previous example. The lead and copper are fixed as sulphides. Wash the aluminium into the ring zone with $0.05M$ hydrochloric acid, dry the filter paper, and punch out the inner spot bearing the mixed sulphide precipitate. Cut out a sector of the ring zone, spray it with a saturated solution of morin in methanol, dry, and rinse in $2M$ hydrochloric acid to remove excess of the reagent and any other morin complexes. Examine the moist sector under ultra-violet radiation. A yellow-green fluorescent ring zone indicates aluminium. Oxidize the fixed material in the small disc over bromine vapour and fume it over ammonia. The disc now contains copper as its tetra-ammine complex and lead as lead sulphate. Place the small disc in the centre of a new filter paper and moisten it with a drop of ethanol. Place the filter paper, together with the disc, upon the ring-oven so that the small disc lies centrally. Wash out all the copper with $3M$ ammonia as if the small disc were a normal sample spot. All the copper is dissolved and collects in the ring zone. Cut out a sector of the filter paper and identify copper as above. The disc now contains only lead, as the sulphate. Add to the disc a drop of a freshly prepared 0.2% aqueous sodium rhodizonate solution and fume it over concentrated hydrochloric acid until the yellow colour of the excess of reagent disappears: a violet-red spot indicates lead.

6.3 SYSTEMATIC SEPARATION SCHEMES

Because it is possible to vary the group precipitation reagents and their physical form (gaseous or liquid), and the washing-out solvents and to combine them in various ways with the different operations, (washing-out, punching, washing into a second filter paper), a large number of ions contained in one sample can be separated into several groups and identified. Care has to be taken that ions collected in particular groups do not interfere with identifications, a condition necessary in any analytical identification scheme.

It should be emphasized that the ring-oven method is *not* paper chroma-tography. In paper chromatography separations are based on the different elution behaviour of the substances, which results in different migration rates and distances from the starting point. In the ring-oven technique the well known methods of separation by precipitation are applied, and the equivalent of the filtration process takes place horizontally.

Several systematic schemes for the analysis of cation mixtures have been developed. A fairly general scheme for 14 metal ions is described here [2]. Only one drop (1-2 μl) containing a few μg of solid sample (depending, of course, on the number, nature and concentration of metals present) is required.

As all the technical operations have already been described, only a brief schematic description is given in Table 6.1.

Table 6.1

Systematic separation scheme using the ring-oven

Place sample drop (in $\leqslant 2M$ HCl) on paper; add H_2S. Wash out with $0.05M$ HCl.

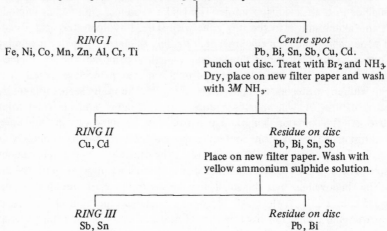

All the ions have been separated into groups, within which individual identifications are possible. Cut out as many sectors from the three rings as are needed for the identification reactions. It is not necessary to detail these reactions at this stage, but there are more than 130 reactions described in the literature for the detection of cations by the ring-oven method [1].

Whereas the separation scheme described above is based on precipitation and filtration, an alternative is based on liquid–liquid extraction [3]; the various extracts are transferred from a capillary-tipped extraction pipette to the centre of a round filter paper and the ions contained in it are washed into the ring zone. It is only necessary to extract very small volumes (1 or 2 drops) of the unknown aqueous solution.

6.4 RING-TO-RING SEPARATION

In some instances it is useful to separate further the substances which have been collected in a ring. There are various possibilities for solving this problem, one being ring-to-ring separation [4].

A special aluminium adaptor is necessary which reduces the size of the hole of the ring-oven [4]; this adaptor (Fig. 6.7) just fits into the hole of the ring-oven and has an internal diameter of 14 mm. Substances can thus be concentrated first in an inner ring zone by use of the adaptor, and after removal of the adaptor can be further separated by use of a solvent which precipitates some of the substances in the inner-ring and transports the others into the outer ring zone. The following example illustrates the application of this principle.

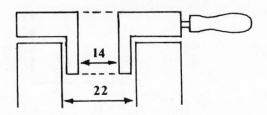

Fig. 6.7 – Aluminium adaptor (dimensions in mm).

Place a drop containing copper and iron in the centre of a filter paper. Wash both ions to the inner ring zone with 0.05M hydrochloric acid. Remove the adaptor, place the filter paper on the ring-oven again and wash it with 3M ammonia containing some ammonium chloride. All the copper goes to the outer ring zone, the iron being precipitated as the hydrated oxide in the inner ring zone. Fume the paper over concentrated hydrochloric acid and spray it with a 1% potassium hexacyanoferrate(II) solution. Two concentric rings will appear, of 14 and 22 mm diameter, consisting of Prussian Blue and red-brown copper hexacyanoferrate(II), respectively.

6.5 ANIONS

The ring-oven method has also been used for the qualitative analysis of anions, and systematic schemes have been proposed for the separation of mixtures of anions [1]. As with cations, the various separations are based on precipitation of certain of the ions at the centre of the paper and washing out the unprecipitated material to the outer ring zone. Many sensitive spot-test reactions for the identification of anions have been adapted for application in the ring-oven method.

6.6 ELECTROGRAPHIC SAMPLING

In the analysis of metallic specimens it is often advantageous to use an electric current for dissolution of the material, rather than an acid or a fusion procedure. The sample is made the anode and a platinum, aluminium or stainless-steel electrode forms the cathode. A piece of filter paper, moistened with a suitable electrolyte, is placed between the sample and anode.

Electrographic sampling [5] is rapid and effective, and has proved useful for investigations of objets d'art, coins, etc., because it is effectively non-destructive. The analysis of a coin serves as a typical example (Fig. 6.8).

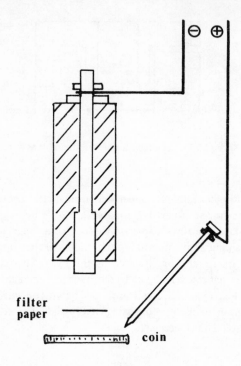

Fig. 6.8 – Electrographic sampling.

Place a small disc of filter paper (e.g. 8 mm in diameter) moistened with
0.1M potassium chloride on the coin to be analysed. Press the cathode, consisting of a metal rod (e.g. stainless steel, 4–5 mm in diameter), contained in an
insulating material, onto the filter paper disc. Bring the anode connection, a
pointed steel probe, into contact with the coin to complete the circuit. As a
source of electricity use a 4.5-V dry battery. Electrolyse for 3–5 sec and then
remove the electrodes. The filter paper disc will contain enough of the metal
ions from the coin for detection.

First fume the disc over hydrochloric acid and then place it at the middle of
a round filter paper. Treat with hydrogen sulphide as described earlier; silver
and copper are precipitated as sulphides. Place the disc on a filter paper on the
ring-oven and wash all the iron, cobalt, nickel and aluminium into the ring-zone
of the filter paper, using 0.05M hydrochloric acid. Oxidize the sulphides in the
disc with bromine and ammonia as described above, place the disc on a new
filter paper, and wash all the copper into the ring zone with 0.05M hydrochloric acid. Place the disc on a third filter paper and transfer all the silver to a
ring zone with 9M ammonia.

On sectors of the three rings, the various ions can be detected, for example, Ag with potassium chromate, Cu with rubeanic acid, Fe with potassium hexacyanoferrate(II), Ni with dimethylglyoxime, Co with 1-nitroso-2-naphthol, and Al with morin.

REFERENCES

[1] Weisz, H., *Microanalysis by the Ring Oven Technique*, 2nd Ed., Pergamon Press, Oxford, 1970.

[2] Weisz, H., *Mikrochim. Acta*, **1954**, 140, 376.

[3] West, P. W. and Mukherji, A. K., *Anal. Chem.*, 1959, **31**, 947.

[4] West, P. W., Llacer, A. J. and Cimerman, C. H., *Mikrochim. Acta*, **1962**, 1165.

[5] Stephen, W. I., *Mikrochim. Acta*, **1956**, 1531.

Appendix

List of Reagents

GENERAL REAGENTS

Compositions expressed in per cent are w/v, and solutions are aqueous, unless otherwise stated.

Acetic acid, $1M$ (57 ml of anhydrous acid per litre).
Acetic acid, $4M$ (230 ml of anhydrous acid per litre).
Ammonia solution, conc. (35%, density 0.88 g/ml), ca. $17M$.
Ammonia solution, $8M$ (500 ml of conc. ammonia solution per litre).
Ammonia solution, $4M$ (225 ml of conc. ammonia solution per litre).
Bromine water (shake water with a little bromine and leave standing in contact with bromine, ca. 3.5% w/w).
Hydrochloric acid, conc. (36% density 1.16 g/ml), ca. $11M$.
Hydrochloric acid, $8M$ (730 ml of conc. acid per litre).
Hydrochloric acid, $4M$ (365 ml of conc. acid per litre).
Hydrogen peroxide solution, 20 volume (ca. 6%).
Hydrogen sulphide in acetone (see p. 236 for preparation).
Nitric acid, conc. (70%, density 1.42 g/ml), ca. $16M$.
Nitric acid, $4M$ (255 ml of conc. acid per litre).
Potassium hydroxide solution, $0.5M$ (28 g/ litre).
Sodium hydroxide solution, $4M$ (160 g/ litre).
Sulphuric acid, conc. (97%, density 1.84 g/ml), ca. $18M$.
Sulphuric acid, $2M$ (111 ml of conc. acid per litre).

OTHER REAGENTS FOR PRELIMINARY TESTS AND CATION SCHEME

Liquids and solutions
Ammonium acetate, $2M$ (15%).
Ammonium acetate, 50%.
Ammonium carbonate, saturated, ca. $10M$.
Ammonium chloride, 20%, ca. $4M$.

Ammoniacal EDTA, 0.5M (18 g of disodium salt with 10 ml of 4M ammonia solution per 100 ml).

Ammonium tetrathiocyanatomercurate(II), (2.7 g of $HgCl_2$ and 3 g of NH_4SCN per 100 ml).

Ammonium oxalate, saturated, ca. 0.3M.

Ammonium thiocyanate, 2M (15%).

Amyl alcohol.

Barium chloride, 0.5M (12% $BaCl_2.2H_2O$).

α-Benzoin oxime (5% in ethanol).

N-Benzoylphenylhydroxylamine (1% in 50% v/v acetic acid).

Butanol/ether (3 vols. of n-butanol with 1 vol. of ether).

Cacotheline (stir 0.25 g with 100 ml of cold water, filter).

Chloroacetate (mix equal volumes of 6M KOH and 1M potassium chloroacetate solutions).

Chloroacetic acid, 4M (38%).

Chloroform.

Cobalt(II) nitrate, 1.5% $Co(NO_3)_2.6H_2O$.

Copper(II) sulphate, 0.1% $CuSO_4.5H_2O$.

Copper uranyl acetate [dissolve 25 g of $Cu(CH_3COO)_2.H_2O$ and 30 g of $UO_2(CH_3COO)_2.2H_2O$ in 450 ml of water and 100 ml of anhydrous acetic acid at 50-60°, cool, add 500 ml of 95% ethanol; let stand 2–3 days and filter if necessary; store in brown bottle].

Diammonium hydrogen phosphate, 10%.

p-Dimethylaminobenzylidenerhodanine (0.03% in acetone).

Dimethylglyoxime (1% in ethanol).

Ethanol.

Glyoxal bis(2-hydroxyanil) (1% in ethanol).

8-Hydroxyquinoline (3% in chloroform).

Iodine, 0.02M (0.25 g of iodine dissolved in 2 ml of 50% KI solution and diluted to 100 ml).

Iodine-starch solution (dilute the 0.02M iodine until it is a very pale yellow and add starch solution. The reagent, which must be freshly prepared, should be dark blue, but should appear transparent in a centrifuge tube).

Iron(II) 2,2′-bipyridyl iodide (dissolve 0.25 g of 2,2′-bipyridyl in 50 ml of water, with warming, cool, add 0.146 g of $FeSO_4.7H_2O$ and shake to dissolve; add 10 g of KI with shaking, let stand overnight, filter and store in a stoppered bottle).

Iron(II) sulphate 1M (28% $FeSO_4.7H_2O$).

Lead acetate, 10% (dissolve 10 g in water, with addition of 4M acetic acid to obtain a clear solution; dilute to 100 ml with water).

Lime water (saturated solution of freshly slaked lime).

Magnesia mixture (dissolve 10 g of $MgCl_2.6H_2O$ and 25 g of NH_4Cl in 100 ml of water; add 35 ml of conc. ammonia, let stand overnight, then filter).

Magnesium chloride, $1.2M$ (24% of $MgCl_2.6H_2O$).

Mercury(II) chloride, saturated, ca. $0.25M$.

Molybdate reagent A [dissolve 34 g of $(NH_4)_6Mo_7O_{24}.4H_2O$ in 70 ml of water and add 30 ml of conc. ammonia].

Molybdate reagent B (dissolve 120 g of NH_4NO_3 in 280 ml of water and add 120 ml of conc. HNO_3).

α-Naphthol (0.5% in $2M$ NaOH).

o-Nitrobenzaldehyde (saturated solution in $2M$ NaOH).

Nitroso-R salt (0.5%).

Oxalic acid, $1M$ [12.6% $(COOH)_2.2H_2O$].

Palladium(II) chloride (dissolve 0.1 g of $PdCl_2$ in 100 ml of water containing 0.5 ml of conc. HCl; store in an amber bottle).

Phenolphthalein (0.1% in ethanol).

Potassium chromate, 5%.

Potassium chromate, 10%.

Potassium hexacyanoferrate(II), 5%.

Potassium iron(III) periodate (dissolve 2 g of potassium metaperiodate, KIO_4, in 10 ml of $2M$ KOH, dilute with water to 50 ml, add 3.5 ml of $0.5M$ $FeCl_3$, dilute to 100 ml with $2M$ KOH).

Potassium iodide, 2%.

Rhodamine B (0.01% in water).

Silver nitrate 1%.

Sodium acetate, $4M$ (54% $CH_3COONa.3H_2O$).

Sodium carbonate, 10% Na_2CO_3.

Sodium chloride, saturated, ca. $6M$.

Sodium hypochlorite, 15%.

Sodium nitrite, 20%.

Sodium nitrite, 2%.

Sodium rhodizonate (mix 1 part of sodium rhodizonate with 20-100 parts of urea; dissolve ca. 10 mg of the mixture in 0.5 ml of water immediately before use).

Solochrome Cyanine (0.1%).

Starch, 1% (triturate 1 g of soluble starch with 2–3 ml of water; stir into 100 ml of boiling water; 'Thyodene' may be used instead).

Sulphanilic acid (1% in $4M$ acetic acid).

Tartaric acid, 30%.

Thiourea, 10%.

Tin(II) chloride, 5% (dissolve 5 g of $SnCl_2.2H_2O$ in 10 ml of conc. HCl; dilute to 100 ml with water).

Titan Yellow, 0.1%.

Zinc chloride, $1M$ (14% $ZnCl_2$).

Zirconium fluoride–Alizarin reagent [dissolve 0.1 g of Alizarin S in 100 ml of water; add 20 drops of 0.2% $ZrO(NO_3)_2$ solution – a red-violet colour develops; add 1% NaF solution drop by drop until the original yellow colour reappears] .

Zirconium nitrate [dissolve with warming 10 g of ZrO $(NO_3)_2$ (Al < 0.1%) in 200 ml of 2*M* HNO_3; add paper pulp to the hot solution, cool and filter through a Whatman No. 40 paper] .

Solids

Aluminium turnings.

Ammonium chloride.

Ammonium peroxodisulphate.

Ammonium thiocyanate.

Borax (powdered).

Calcium fluoride.

Chalk (precipitated).

Devarda's alloy.

Iron(II) sulphate (hydrated).

Iron wire.

Lead acetate paper.

Litmus paper.

Microcosmic salt (ammonium sodium hydrogen phosphate).

Potassium hexacyanoferrate(III).

Potassium nitrate.

Silica (powdered) or sodium silicate.

Sodium bismuthate.

Sodium carbonate (anhydrous).

Sodium fluoride.

Sodium thiosulphate (hydrated).

Starch-iodide paper.

PREPARATION OF H₂S–ACETONE REAGENT

Pass hydrogen sulphide from a conventional Kipp's generator through a flask containing distilled water, then into a drying tower containing granulated silica gel (*not* impregnated with a cobalt salt) or anhydrous calcium chloride (the first dessicant is preferred; the calcium chloride soon forms a hard cake of the hydrated salt and impairs the flow of gas). Lead the dried gas into a gas absorber filled with 500 ml of acetone previously dried with anhydrous sodium sulphate, and placed in a salt-ice cooling mixture. Pass a steady stream of gas (the maximum output from a generator) for about two hours or until the solution is almost saturated, as indicated by the rate at which acetone is drawn back into the system when the gas supply is cut off. Transfer the solution to a stoppered reagent bottle. The gas may also be bubbled directly into acetone through a glass delivery tube, but the absorption of the gas is less effective and saturation takes longer to attain. For use in the laboratory, transfer the solution to 60-ml reagent bottles of the ground-glass dropping-pipette type fitted with vulcanized polythene teats. The ground joints can be lightly smeared with silicone grease to prevent undue evaporation of the reagent.

Note
The difficulties which have been experienced when using this reagent in climates where the temperature can exceed 30° can be overcome by substituting methyl ethyl ketone for acetone when making up the reagent.

ADDITIONAL REAGENTS FOR THE ANION SCHEME, COMBINATIONS OF ANIONS AND PARTICULAR SUBSTANCES

Solutions
Acetate buffer (dilute 50 ml of $2M$ acetic acid plus 25 ml of $2M$ NaOH to 100 ml).
Acetic acid, $0.88M$ (5% v/v).
Ammonium molybdate, 10%.
Ammonium sulphide.
Barium chloride, saturated, ca. $1.4M$.
Barium chloride/sodium nitrite (0.5% $BaCl_2$. $2H_2O$, 5% $NaNO_2$).
Barium nitrate, $0.25M$ (6.5%).
Cadmium sulphate, 10%.
Calcium nitrate, $1M$ (dissolve 10 g of $CaCO_3$ in conc. HNO_3 and dilute to 100 ml).
Copper acetate [dissolve with warming 1 g of $Cu(CH_3COO)_2.H_2O$ in 30 ml of water containing 1-2 ml of $4M$ acetic acid].
Denigès reagent (dissolve 5 g of HgO in 20 ml of conc. H_2SO_4; add 80 ml of water).
p-Dimethylaminobenzaldehyde (dissolve 5 g in toluene, add 20 g of trichloroacetic acid, make up to 100 ml with toluene).
Fluorescein (0.5% in ethanol).
Gelatin, 1%.
Iodide–iodate (2.5% of KI and 2.5% KIO_3; *must be freshly prepared*).
Iodine, $0.01M$ (0.13 g of iodine dissolved in 1 ml of 50% KI solution and diluted to 100 ml).
Iron(III) ammonium sulphate, 1% [carefully neutralize any excess of acid, but avoid precipitation of hydrated iron(III) oxide].
Lanthanum nitrate, 5%.
Lead nitrate, $1M$ [33% $Pb(NO_3)_2$].

Le Rosen's reagent (10 volumes of conc. H_2SO_4 with 1 volume of 37% aqueous formaldehyde, freshly prepared).

Mercury

Methylene Blue (0.03%).

Nickel–dimethylglyoxime (add, with stirring, a solution of 2.8 g of dimethyl-glyoxime in 300 ml of ethanol to a solution of 2.3 g of $NiSO_4.7H_2O$ in 300 ml of water; set aside for 30 minutes, then filter.

Perchloric acid, $4M$ (2 volumes of 60% $HClO_4$ and 1 volume of water).

Phosphoric acid, conc. (90%, density 1.75 g/ml).

Potassium permanganate, $0.02M$ (0.32%).

Salicylaldehyde (boil 1 g of salicylaldehyde with 120 ml of water and 2 ml of acetic acid until the oil disappears; set aside for 24 hours, then filter).

Silver ammine solution A (17 g of $AgNO_3$ in 30 ml of water).

Silver ammine solution B [25 ml of saturated $(NH_4)_2CO_3$ solution with 10 ml of conc. ammonia and 100 ml of water].

Silver periodate reagent (25 ml of 2% KIO_3 solution with 2 ml of conc. HNO_3 and 2 ml of 10% $AgNO_3$ solution).

Sodium thiosulphate, 10% $Na_2S_2O_3.5H_2O$.

Titanium(IV) sulphate, 1% (fuse 1.7 g of TiO_2 with $KHSO_4$, cool and dissolve in 100 ml of cold $2M$ H_2SO_4).

Toluene

Zinc sulphate, saturated, ca. $3.3M$.

Solids

Ammonium nitrate.

Diphenylamine.

Lead dioxide.

Potassium permanganate.

Sodium nitrite.

Zinc dust.

ADDITIONAL REAGENTS FOR LESS COMMON ELEMENTS AND INSOLUBLE SUBSTANCES

Solutions and liquids

Acetylacetone (50% in toluene).

Alizarin (0.1% in conc. H_2SO_4).

Ammonium molybdate–nitric acid (10% solution made acidic with $4M$ HNO_3).

Ammonium thiosulphate, 50%.

2,2'-Bipyridyl, 2%.

Disodium hydrogen phosphate, 10%.

Iron(III) chloride, $1M$ (16.2% $FeCl_3$).

Mandelic acid, 10%.

p-Nitrobenzeneazo-orcinol (0.025% in $1M$ NaOH).

Potassium tetracyanonickelate (dissolve 1 g of $NiSO_4.6H_2O$ in 10 ml of water and add 10 ml of 10% KCN solution. Filter and wash. Transfer the precipitate to a beaker with 20 ml of water. Boil and add 10% KCN solution until the precipitate has *almost* dissolved. Cool and filter).

Tin(II) chloride, 10% (dissolve 10 g of $SnCl_2.2H_2O$ in 38 ml of conc. HCl and dilute to 100 ml with water).

o-Tolidine (1% in ethanol).

Zirconium nitrate–Alizarin reagent [dissolve 0.05 g of $ZrO(NO_3)_2$ in 40 ml of water and 10 ml of conc. HCl and add to a solution of 0.05 g of Alizarin S in 50 ml of water. Mix].

Solids
Hydroxylammonium sulphate.
Iron(III) ammonium sulphate.
Oxalic acid.
Potassium hydrogen sulphate.

Index

Reagents are listed individually in the Appendix.

A

acetate, 31, 137, 168, 170
  complexes, 51
acetic acid, 45, 104, 138
acetylacetone, 79, 85
acid, Brønsted, 29
acid–base dissociation constants, 39, 40
  indicators, 43
  pH, 40
  polyprotic, 40
  reactions, 29, 38
activity, 37
  coefficient, 37
aging of sulphide precipitates, 110
Alizarin S, test for borate, 217
alpha-coefficients, 56
  and redox, 69
aluminium, 164, 166, 168, 187, 209, 227, 228, 230
  hydroxide, K_{sp}, 120
  hydroxo-complexes, 123
  8–hydroxyquinoline complex, 31, 70, 72
  oxide, 143, 213
ammine complexes, 50, 52, 76, 85, 121, 131
  stability constants, 122
ammonia, 30, 39, 46, 167, 171
ammonium, 164, 165, 171
  carbamate, 131
  polysulphide and Copper–Tin Group, 116
amyl acetate, 34
  alcohol, 34
anions, interference in cation separation, 127, 167, 170, 191
  less common, 210
  organic, 128, 170
  removal, 178, 191
  tests, 137, 167, 193, 197
anion separation scheme, 128, 191, 192, 211, 228
antimony, 108, 164, 165, 166, 185, 186, 228

antimony (*contd.*)
  oxide chloride, 99, 174
  sulphide, in Silver Group, 99
  K_{sp}, 110
  thio-salts, 99
apparatus, 147
aqua regia, 126, 138, 175
arsenate, 133, 135, 166, 172, 185, 202, 207
arsenic, 108, 165, 171, 185, 186, 202, 207
  reduction, 111
  sulphide in Copper–Tin Group, 110
    in Silver Group, 99
  thio-salts, 99
arsenite, 131, 133, 172

B

barium, 165, 182
  chromate, 104, 105, 107, 108, 181
  EDTA complex, 106
  flame test, 105, 182
  sulphate, 32, 54, 101, 102, 139, 213
    metathesis with carbonate, 103, 129
  tests, 182
base, Brønsted, 29
  dissociation constant, 39
beaker, 148
benzoate, 133, 195
benzoic acid, 70, 177
  in Silver Group, 99
  interference, 128
α–benzoin oxime test for copper, 184, 234
N–benzoylphenylhydroxylamine test for tin, 185, 234
Bergman, 18, 20
beryllium, 119, 123, 124, 203, 209
  carbonate complex, 134
  hydroxide, 119, 123
  K_{sp}, 120
  hydroxo complex, 123, 124, 210
  test, 209

biacetyl, 62
bicarbonate, *see* hydrogen carbonate
2,2'-bipyridyl, 87, 216
   reagent for cadmium, 184
bismuth, 108, 166, 183, 228
   oxide chloride, 99, 179
   sulphide, K_{sp}, 110
   tests, 183
blowpipe, 22, 23, 165
borate, 120, 165, 166, 168, 171, 217
   interference, removal of, 128, 178
borax bead test, 166
boric acid, in Silver Group, 99
   interference, 128
boron nitride, 141, 143, 213, 215
Brilliant Green, 79
bromate, 133, 164, 165, 169, 195, 212, 217
bromide, 131, 133, 165, 168, 169, 174, 194
bromine, test for, 168
Bromothymol Blue, 44
Brønsted acids and bases, 29
buffer capacity, 47
   index, 47
   solutions, 46

C

cacotheline, reagent for tin, 185
cadmium, 108, 165, 184, 228
   ammine, 118
   chloro complexes, 49, 112, 118
   separation from copper, 118
   sulphide, K_{sp}, 110, 111, 125
      precipitation, 111, 125
calcium, 165, 180, 181, 214
   fluoride, 139, 213, 236
   Group, 95, 101, 176, 204
      in Silver Group, 99
      separation, 101, 103, 180, 204, 206
   oxalate, 56, 80
   salts, K_{sp}, 135
   sulphate, 101, 102, 181
   tartrate, 61
carbamate, 131
carbon, 213, 214
   dioxide test, 156, 168
   monoxide test, 170
carbonate, 129, 133, 137, 164, 168, 197
   complexes, 130
   metathesis with sulphate, 103
catalysed reactions, 33
cation separation scheme, 95, 174, 176, 227
cations, less common, 203
centrifuge, 155
   tubes, 149

cerium, 65, 67, 119, 124, 208, 210
cerium(III) hydroxide, 119, 123, 124
   K_{sp}, 120
cerium(III) oxalate, 124
cerium(IV) hydroxide, 124
charcoal block, 22, 165, 214
chelates, 31
   complexes, 31
   ring size, 84
chemical equilibria, 34
chlorate, 164, 169, 196
chloride, 96, 131, 133, 165, 169, 173, 194
   complexes, 55, 118
   precipitates, K_{sp}, 97
chlorine dioxide, 169
chlorine, test for, 168, 171
chlorine water, 201
chloroacetate dissolution of Tin Group, 116
chloroacetic acid, 111
4-chloro-4'-ammoniumbiphenyl, 33, 79
chromate, 31, 133, 136, 164, 167, 174, 196, 200, 213
   separation of Ba, Sr, Pb, 104, 108
chromic acid, 104
chromium, 119, 123, 164, 166, 187, 210, 214, 228
   ammine complexes, 121
   hydroxide, 119, 123
      K_{sp}, 120
   oxide, 139, 143, 213, 214
chromyl chloride, 169
citrate, 120, 169, 197
   interference, 128
cobalt, 32, 124, 164, 166, 189, 228, 230
   ammine complexes, 52, 122
   chloride complex, 36
   co-precipitation, 122
   glass, 159
   hydroxide, K_{sp}, 120
   sulphide, K_{sp}, 110, 125
      precipitation, 124
colloidal suspension, 63
colloids, flocculation, 64
   lyophilic, 64
   lyophobic, 64
   peptization, 64
colour, 164, 214
   formation, 79
common-ion effect, 54
complexation reactions, 30, 48, 55, 80
   and extraction, 74
   and masking, 55
   and pH effect, 80
   and redox, 68
   stability of complexes, 49, 83

complexes, acetate, 51
ammine, 50, 76, 85, 121, 122
carbonate, 130
chelate, 31
chloride, 55
cyanide, 50, 77
EDTA, 81
ethylenediamine, 85
formation constants, 49
alpha-coefficient, 56
conditional, 57
overall, 50
side-reaction coefficient, 49
stepwise, 49
hydroxide, 51
8-hydroxyquinoline, 73, 74, 81
inert, 52
labile, 52
mixed metal, 78
organic reagent, 79, 80
and pH, 80
oxalate, 55
complexing agents, 83
conditional concentration, 57
constants, 57
potential, 68
copper, 108, 164, 165, 166, 184, 214, 226, 227, 228, 230
ammine complex, 118
chloride, 97
complexes, 113
Group separation, 117, 182, 183, 206
hexacyanoferrate(II), 139, 142, 213, 214
8-hydroxyquinoline complex, 80
iodide in Silver Group, 99
separation from cadmium, 118
sulphide, 108
K_{sp}, 110
uranyl acetate reagent, 190
Copper-Tin Group, 95, 108, 176, 203
separation, 115, 182, 186, 204, 206, 208
Co-precipitation, 61
bulk, 61
in Copper-Tin Group, 114
in Iron Group, 122, 188
inclusion, 61
surface, 61
corrosive sublimate, 18
crucible, 148
crystal growth, 60
cudbear, 19
cyanate, 137, 168, 197, 198
cyanide, 137, 164, 165, 169, 172
complexes, 50, 77
destruction, 78, 191

D

Debye–Hückel equation, 38
demasking, 76
cyanide complexes, 78
EDTA complexes, 78
fluoride complexes, 78
Denigès reagent, test for citrate, 197
destruction of interfering ions, 191
Devarda's alloy, 172, 197
diammine silver reagent, 30, 50, 130, 131, 133, 192
dichromate, 31, 164, 167, 169, 200
diethyldithiocarbamate, 33
p-dimethylaminobenzaldehyde, test for succinate, 197
p-dimethylaminobenzylidenethodamine, reagent for mercury, 180, 183
for silver, 180
dimethylglyoxime, 32, 87, 214
reagent for bismuth, 183
for nickel, 189
diphenylamine, reagent for oxalate, 196
dissociation constant, acid, 39
base, 39
of water, 41
distribution coefficient, 71
of pH, 72
dithionate, 199
dithionite, 167, 168
dry tests, 164

E

EDTA, 40
complexes, 81
dissolution of sulphate precipitates, 105
electrode potential, 65
conditional, 68
effect of masking, 79
of pH, 68
formal, 68
standard, 65
electrographic sampling, 229
Epsom salts, 18
equilibrium, 34, 35
constant, 35
effect of solvent, 36
of temperature, 36
multiple, 75
partition, 70
redox, 66, 67
ethylenediamine complexes, 85
evaporation, 160
extraction, 70
formation of complexes, 79
nature of, 73

F

Fajans's rules, 84
ferricyanide, *see* hexacyanoferrate(III)
ferrocyanide, *see* hexacyanoferrate(II)
flame tests, 24, 87, 159, 165, 202, 214
fluorescein paper, test for bromate, 194
fluoride, 120, 133, 135, 158, 169, 171, 196, 214
  interference, 127, 178
  lead plate test, 158, 171
  removal of, 178
fluorides, insoluble, 141, 142, 213, 214
fluorosilicate, 158, *see* also hexafluorosilicate
formal electrode potential, 68
formate, 169, 172, 196
formation constant, 49
fuming, 160
functional groups, 87
fusion, 139, 140, 142, 143, 217
  apparatus, 148, 149
  potassium hydrogen sulphate, 140, 143
  sodium carbonate, 140, 142
  sodium peroxide, 140, 142

G

gall test, 19
gas generator for ring-oven, 225
gases, precipitants, 224
  testing, 156, 164, 165, 167, 168, 169
  toxic, 163
Glauber's salt, 18
glyoxal-bis(2-hydroxyanil), reagent for calcium, 181
group separation, anion, 128, 191, 192, 211, 228
  cation, 95, 174, 176, 277
  ring-oven, 227, 228
Groups, 95, 176, 203
  Calcium, 101, 180
  Copper, 115, 183, 206
  Copper–Tin, 108, 182, 186
  Iron, 119, 187, 208
  Magnesium, 126, 190
  Silver, 96, 179, 205
  Tin, 115, 184, 207
  Zinc, 124, 188
Guldberg and Waage's law, 35
  applications, 39

H

heat, effect of, 164
heating block, 151
Henderson equation, 47
hepatic air, 21

hexacyanoferrate(II), 131, 133, 169, 192, 193, 216
hexacyanoferrate (III), 52, 131, 169, 192, 193, 216
hexafluorosilicate, 159, 165, 169, 171
  destruction of, 191
  lead plate test, 158, 159
hexamethylene tetramine, 62
hydrazine, test with salicylaldehyde, 34, 80, 201
hydrofluoric acid, 33
hydrogen carbonate, 137, 164, 197, 198
  peroxide, 118, 187
    test for, 202
  sulphate, 102
  sulphide, equilibrium, 109
    reagent, 108, 154
    test for, 167
hydroxides, 51, 168
  K_{sp}, 120
hydroxo complexes, 51
hydroxylamine, test for, 201
8-hydroxyquinoline complexes, 31, 58, 72, 73, 74, 80, 81, 83, 126
hypobromite, 168
hypochlorite, 118, 168, 201
  destruction of, 191
hypophosphite, 164, 169, 200

I

indicators, acid-base, 43
  table, 44
  universal, 44
inert complexes, 52
insoluble substances, 138, 141, 213
  charcoal block reactions, 165
  fusion processes, 139
interfering anions, 120, 191
  destruction, 191
iodate, 133, 164, 165, 195, 212
iodide, 111, 119, 126, 131, 133, 164, 165, 169, 174, 194
  and copper(II), 99
  and iodate solution, reagent for dichromate, 200
ionic product, and K_{sp}, 53
  water, 41
ionic strength, 37
iron, 166, 168, 187, 228, 230
iron(II), 164
  ammine complexes, 122
  hydroxide, K_{sp}, 120
  sulphide, 126
    K_{sp}, 110

iron(III), 164, 168, 187, 223
  hydroxide, 123
  K_{sp}, 120
  oxalate complex, 124
  oxide, 139, 213, 214
  vanadate, 122
Iron Group, 25, 119, 187, 203
  complex formation, 121
  interferences, 120
  separation, 122, 187, 204, 208

K
Kipp's apparatus, 22
Klaproth, 19

L
labile complexes, 52
law of mass action, 35
lead, 96, 165, 166, 179, 205, 228
  acetate, 51
  chloride, 96, 97
    complexes, 97
  chromate, 106, 108, 141, 142, 213, 214
  EDTA complex, 106
  in Calcium Group, 101
  in Silver Group, 96
  oxalate, 81
  plate, 158
  sulphate, 100, 101, 102, 141, 213
  sulphide, 104
    K_{sp}, 110
lead(IV) oxide, 194
Le Rosen's reagent, test for benzoate, 195
ligands, 83
  and stability constants, 83
  basicity, 83
lithium, 126, 165, 190
litmus, 19

M
magnesia mixture, 185
magnesium, 166, 190
  ammine complex, 122
  ammonium arsenate, 118
  EDTA complex, 107
  Group, 95, 126, 176, 191
    separation, 126, 190
  hydroxide, K_{sp}, 120
  8-hydroxyquinoline complex, 126
  tellurate, 118
mandelic acid, reagent for zirconium, 210

manganese, 164, 166, 189, 228
  ammine complexes, 122
  hydroxide, K_{sp}, 120
  oxalate, complex, 124
  oxide, 217
  sulphide, 124, 125
    K_{sp}, 110
MAQA, 13
  scheme, 94
masking, 76, 78, 81
  and temperature, 78
  kinetic, 78
Meissner test, 214
mercury(I), 96, 164, 179, 205
  chloride, 96, 100
mercury(II), 108, 164, 179, 183
  aminochloride, 118
  chloride complexes, 113
  iodide, 55
  sulphide, 110, 114, 117
    dissolution, 115, 116, 117
    K_{sp}, 110
metathesis, of sulphate precipitate, 103
  with sodium carbonate solution, 103, 129
Methyl Orange, 44
  Red, 43
Methylene Blue, 196
microburner, 152
microcosmic salt bead, 166, 214
mixed metal complexes, 78
molybdate, 133, 135, 136, 203, 210, 213
molybdenum, 108, 113, 207, 213, 216
  oxide, 141, 213
  sulphide, 110, 113
molybdoarsenic acid, 119
molybdoperiodate, 77
morin, reagent for aluminium, 227

N
Nernst equation, 65
nickel, 124, 164, 166, 189, 226, 228, 230
  ammine complexes, 122
  cyanide complex, 30, 50
  dimethylglyoximate, 31, 55, 62
  hydroxide, 120
  spoon, 147, 148
  sulphide, 124
    K_{sp}, 110
nickel–dimethylglyoxime, reagent for chromate, 200
nitrate, 137, 164, 165, 169, 172, 197
  test in presence of nitrite, 197
nitride, 167

nitrite, 168, 172
p-nitrobenzeneazo-orcinol, reagent for
  beryllium, 209
nitroso-R salt, 79
  reagent for cobalt, 189
nitrous fumes, 168
nucleation, 59

O

oil of vitriol, 18
organic acids, 164, 168, 170
  complexes, stability constants, 83
  interference, 127
    removal of, 178
organic reagents, 78, 79, 80
  and masking, 83
  synthetic reactions, 34
Ostwald ripening, 63
overall stability constants, 50
oxalate, 133, 135, 164, 169, 196
  complexes, 55
  interference, 128
oxalates, 51, 81, 120
  pH effect on, 58
oxalic acid, 40
oxides, 164
  insoluble, 213
oxidizing agents, 168, 171
oxygen, test for, 167
ozone, test for, 167

P

partition coefficient, 71
  equilibrium, 70
peptization, 64
perchlorate, 196
periodate, masking with molybdate, 77
  separation, 131
permanganate, 137, 167
  destruction of, 191
peroxides, 164, 167, 168
peroxodisulphate, 137, 167, 168, 169,
  189, 202
  destruction of, 191
pH, 41
  and organic reagents, 80
  effect on solubility, 56
  of buffers, 56
1,10-phenanthroline, 79, 85, 87
phenolphthalein, 44
phosphate, 120, 133, 135, 166, 168
  interference, 127
  removal of, 127, 178
phosphide, 167
phosphine, 167

phosphite, 164, 169, 200
phosphoric acid, 41
pipette, 150
  self-filling, 225
polymerization reactions, 31
polyphosphates, 31
polysulphide, 168
polythionate, 200
post-precipitation, 62
  of Zinc Group, 114
  of zinc sulphide, 114
potassium, 126, 165, 172, 202
  hydrogen sulphate (pyrosulphate) fu-
    sion, 140, 143, 217
precipitate transfer, 156
  washing, 155
precipitation, aging, 62
  completeness, 61
  from homogeneous solution, 62
  induction period, 60
  nucleation, 59
  process, 59
  reactions, 32, 52
    common-ion effect, 54
    co-precipitation, 61
    induced, 60
    salt effects, 59
  solubility, 54
    and organic solvents, 54
    and pH, 56
    and salt effects, 59
    colloidal, 63
    common-ion effect, 54
    complexation, 55
    intrinsic, 53, 82
    molecular, 54, 55
preliminary tests, dry, 163, 164
protonation constant, 39
Prussian Blue, 19
prussiate of potash, 19
pyrolusite, 217

R

reactions, see individual types, e.g.
  acid-base
reagent bottles, 153
reagents, 233
redox equilibria, 65
  and complexation, 68
redox reactions, 32, 64
  electrode potentials, 65
Rhodamine B, reagent for antimony, 185
ring-oven, 221
  electrographic sampling, 229
  gas generator, 225
  separations, 224, 228, 229

ripening, internal, 63
  Ostwald, 63
rocks, analysis, 17, 18, 19

S

salicylaldehyde test for hydrazine, 34, 80, 201
salicylate, 133, 195
salicylic acid, 177
  in Silver Group, 99
salting-out, 75
salts, 45
  insoluble, 213
  of weak acids and bases, 45
saturated calomel electrode, 66
selenate, 108, 113, 117, 133, 203, 210, 212
  flame test, 202
selenite, 108, 113, 117, 131, 133, 203, 210, 212
  flame test, 202, 212
selenium, 203, 206, 207
  in Copper-Tin Group, 140, 142
sensitivity, 25
side-reaction coefficient, 56
silica, 33, 139, 142, 195, 213, 214
silicate, 120, 133, 139, 159, 166, 171, 213, 214, 215
  in Silver Group, 99, 177, 178
  interference, 127
    removal of, 178
  lead plate test, 158, 171, 195
silicon, 141, 213, 214, 215, 217
  carbide, 143, 213, 214
silver, 96, 116, 180, 230
  ammine complex, 50, 76, 100
  as anion group-reagent, 131, 192
  bromide, 139, 214
  chloride, 96, 100, 139
  chromate, 134
  Group, 95, 96, 176, 179, 203
    separation, 100, 179, 205
  interferences, 99
  iodide, 100, 139, 214
  oxalate, 134
  salts, insoluble, 214
  K_{sp}, 133
  thiocyanate, 139, 213
sodium, 126, 165, 190
  bismuthate, 189
  carbonate, metathesis, 103, 129
    extract, 103, 130, 171, 202, 216
    fusion, 140, 142, 217
  hexanitrocobaltate(III), reagent for potassium, 172
  hydroxide extract, 215

sodium (*contd.*)
  peroxide fusion, 140, 142
Solochrome Cyanine, reagent for aluminium, 187
solubility, intrinsic, 53, 82
  molecular, 54
  of sulphates, 103
  product, 53
    aging effect, 110
    complexation effect on, 81, 111
    conditional, 57, 81
    of calcium salts, 135
    of chlorides, 97
    of chromates, 108
    of hydroxides, 120
    of 8-hydroxyquinolinates, 83
    of ion-pairs, 74
    of oxalates, 58
    of silver salts, 133
    of sulphates, 102
    of sulphides, 110
  tests, 167
solvation, 75
solvent extraction, 70
spatula, 147, 148
spirit of wine, 18
spoon, 147, 148
stability constants, 49
  conditional, 57
  of ammines, 85, 122
  of ethylenediamine complexes, 85
  of hydroxide complexes, 52
  of organic complexes, 83
  overall, 53
  stepwise, 49
starch solution, reagent for iodine, 194
stirring rods, 149
strontium, 165, 181
  chromate, 104, 105, 108
  EDTA complex, 106
  flame test, 165, 182
  8-hydroxyquinoline complex, 74
  sulphate, 101, 102, 139, 213
sublimates, 164
succinate, 197
sulphamates, test for, 201
sulphamic acid, 62
sulphate, 101, 103, 133, 135, 136, 165, 166, 173, 196, 213
  precipitate, metathesis, 103
sulphates, insoluble, 166, 213
  K_{sp}, 102
sulphide, 131, 133, 164, 167, 168, 193, 198
sulphides, 108, 124
  mechanism of precipitation, 114
  solubility products, 110

sulphite, 137, 165, 168, 198, 199
sulphur dioxide, test for, 167
supersaturation, 59, 60
suspension, colloidal, 63
syrup of violets, 18

T
tartrate, 51, 120, 135, 164, 169, 196
    interference, 128
teat pipette, 150
tellurate, 113, 131, 135, 208, 210, 212, 213
tellurite, 108, 113, 131, 208, 210, 212
tellurium, 203, 208
    in Copper-Tin Group, 111, 117, 119
    flame test, 202
tetraphenylborate, 80
thallium, 202, 205
thallium(I) chloride, 96, 99
    flame test, 202
    sulphide, 125, 126
    K_{sp}, 126
thio-anions, 99, 115
thiocyanate, 131, 133, 169, 192, 194
thiodiglycollate, 115
thiosulphate, 137, 168, 198, 199
    destruction of, 191
    interference, 127
thiourea, reagent for bismuth, 183
thorium, 119, 124, 210
    hydroxide, 119, 123
    K_{sp}, 120
    oxalate, 55, 124
    complexes, 124
Thymol Blue, 44
thymolphthalein, 44
Thyodene, 195
tin, 108, 165, 166, 185, 186, 214, 228
Tin Group, dissolution of sulphides, 115, 116
    separation, 118, 119, 184, 207, 208
tin(IV) oxide, 139, 143, 213
    sulphide, 110, 111
tin(II) sulphide, K_{sp}, 110
Titan Yellow, reagent for magnesium, 190
titanium, 143, 203, 208, 214, 216, 228
    hydroxide, 119, 123, 203
    K_{sp}, 120
    oxide, 139, 143, 213
    phosphate, 123, 139, 213
o-tolidine, 216
tongs, 149
toxic gases, 163
trimetaphosphate, 31
tripolyphosphate, 31

tungstate, 96, 133, 203, 210, 212
    in Silver Group, 99
tungsten, 96, 141, 143, 205, 214, 216
    metal, 213, 214
    oxide, 100, 141, 213, 214
    separation, 100

U
uranate, 210, 212
uranium, 119, 124, 203, 210, 212
    carbonate complex, 134
    peroxouranate, 124
    precipitation as diuranate, 123, 203
uranyl hydroxide, 119
    K_{sp}, 120

V
vanadate, 122, 133, 136, 203, 210, 213
vanadium, 119, 123, 203, 209
vanadium(V), 122, 136, 203, 210
vanadium(IV) hydroxide, K_{sp}, 120
vegetable tinctures, 19
volatilization reactions, 33

W
washing precipitates, 155
water, dissociation, 41
    dissolution of sample, 167, 174
    mineral, analysis, 20

Z
zinc, 124, 165, 166, 168, 188, 228
    ammine complexes, 122
    EDTA complex, 108
    Group, 95, 124, 176, 188
        co-precipitation with Iron Group, 122, 188
        post-precipitation in Copper-Tin Group, 114
        separation, 126, 188, 204
    hydroxide K_{sp}, 120
    8-hydroxyquinoline complex, 55, 74
    sulphide, 114, 124
        K_{sp}, 110
zirconium, 119, 123, 143, 209, 210, 216
    hydroxide, 119
        K_{sp}, 120
    oxide, 139, 143, 213
    phosphate, 123, 139, 213